Hybrid Electric Vehicle Design and Control

About the Authors

Yangsheng Xu is Chair Professor and Pro-Vice Chancellor of the Chinese University of Hong Kong. He is Professor of Automation and Computer-Aided Engineering, specializing in Robotics, Dynamics and Control, and Manufacturing.

Jingyu Yan is Research Associate of the Chinese University of Hong Kong, specializing in Electric Vehicles, Battery Management System, and Predictive Control.

Huihuan Qian is Research Assistant Professor of the Chinese University of Hong Kong, specializing in Robotics, Dynamics and Control, and Omnidirectional Vehicles.

Tin Lun Lam is Research Associate of the Chinese University of Hong Kong, specializing in Robotics, Human-Machine Interface, and Intelligent Control.

Hybrid Electric Vehicle Design and Control

Intelligent Omnidirectional Hybrids

Yangsheng Xu

Jingyu Yan

Huihuan Qian

Tin Lun Lam

McGraw Hill Education

New York Chicago San Francisco
Athens London Madrid Mexico City Milan
New Delhi Singapore Sydney Toronto

Cataloging-in-Publication Data is on file with the Library of Congress.

Hybrid Electric Vehicle Design and Control: Intelligent Omnidirectional Hybrids

1 2 3 4 5 6 7 8 9 0 CTPS/CTPS 1 2 0 9 8 7 6 5 4 3

ISBN 978-0-07-182683-9
MHID 0-07-182683-1

The pages within this book were printed on acid-free paper.

Sponsoring Editor
Michael Penn

Acquisitions Coordinator
Amy Stonebraker

Editorial Supervisor
David E. Fogarty

Project Manager
Anupriya Tyagi, Cenveo Publisher Services

Copy Editor
Patti Scott

Proofreader
Namita Panda

Indexer
Robert Swanson

Production Supervisor
Richard C. Ruzycka

Composition
Cenveo Publisher Services

Art Director, Cover
Jeff Weeks

Contents at a Glance

Contents

Preface

Intelligent omnidirectional hybrid electric vehicle (HEV) has been a challenging topic in engineering that has gained much attention from researchers. Literature has reported successful implementation of HEVs. However, there are seldom electric vehicles designed to be capable of four-wheel independent steering so that they can move omnidirectionally. The authors of this book have been actively contributing to the field of omnidirectional HEVs. This book organizes the most significant results from the authors on the topic and serves as a complete collection of materials about design and control of intelligent omnidirectional HEVs.

In this book a comprehensive study and analysis of the omnidirectional HEV is presented, including system design and implementation, the control of four-wheel independent driving and steering, the battery management system, and the energy management system. The scope of this book is very broad, from hardware such as mechanical and electrical implementation to software such as intelligent control algorithms. It covers all important aspects and each chapter focuses on a specific issue and is self-contained. That means you can select the appropriate chapter which suits your particular need. Therefore, it is a valuable reference for vehicle engineers to understand every critical part of the omnidirectional HEV.

This book originates from the papers and theses of the authors. It presents the novel design and implementation of the omnidirectional HEVs with detailed experimental result and analysis. In addition, some of the novel techniques mentioned in the book are not only limited to HEVs or four-wheel steering vehicles, but can also be applied on the other types of vehicles. For instance, the power management system based on model predictive control can be applied to any electric vehicles.

This book is appropriate for postgraduate students, research scientists, and engineers with interests in the intelligent omnidirectional HEV. In particular, the book will be an excellent reference for those interested in the topics of mechanical design, implementation, and control of omnidirectional HEVs.

Hybrid Electric Vehicle Design and Control

CHAPTER 1

Introduction

1.1 Background

Transportation has been one of the most essential parts of our society since the beginning of human civilization. Automobiles, defined as wheeled motor vehicles used for transporting passengers, are undeniably the most important transport vehicles in our daily lives. Nevertheless, the tremendous increase in quantities of automobiles in recent years has caused a number of problems for society. The problems can basically be categorized as follows:

Energy and environmental problems. Since the market of all types of automobiles has seen a boom over the last several decades, energy shortage and environmental disruption have become the most serious worldwide problems. In America, for example, road vehicles are one of the most important pollution sources, emitting 18 percent of suspended particulates, 27 percent of volatile organic compounds, 28 percent of Pb, 32 percent of nitrogen oxides, 62 percent of CO, and 25 percent of CO_2.[18] Meanwhile, America's petroleum consumption for transportation accounts for 71 percent of total petroleum consumption and 95 percent of total transportation demand, as shown in Fig. 1.1.[110] It is undoubtedly time, therefore, to change the propulsion source of the traditional internal combustion engine (ICE) vehicle.

Efficiency problem. With the fast growth of vehicles in urban areas, parking has become a significant problem in the design and development of cities. Crowded car parks are

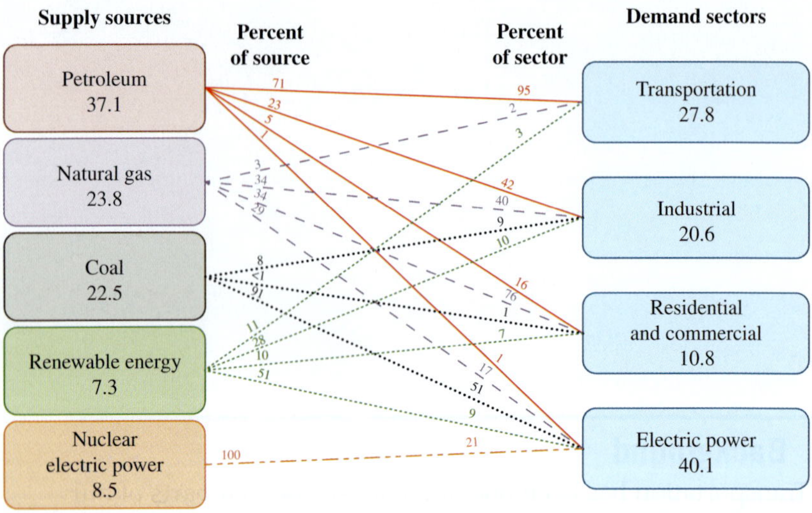

FIGURE 1.1 U.S. primary energy consumption by source and sector (quadrillion Btu), 2008.

common in most of the major cities in the world. Space is badly needed not only for vehicle docking, but also for the room allowing vehicles to move in and out. If the supply of land cannot match such a high demand, this will eventually lower the whole city's efficiency and living standard as well. For example, much more time is required for a driver to search for a vacant parking space near his or her work place.

To tackle the above problems, one naive way is to limit the number of circulating vehicles. This may be one of the short-term solutions; however, the fundamental problem still exists. Some researchers search for the solution in new technology. For example, to solve the energy and pollution problems, electric vehicles (EVs) are recognized as the most accessible way of replacing traditional-fuel vehicles, due to their promising ability to reduce energy consumption and emissions simultaneously. Hence, EVs have attracted great deal of attention from both academia and the automobile industry.

One popular type of EV is the *hybrid electric vehicle* (HEV), which is defined as a vehicle in which propulsion energies are available from two or more types of energy stores, source, or converters and at least one of them can deliver electrical energy. One of the advantages of HEV is that it could combine the benefits from more than one propulsion source. For example, since electric traction motors provide quick acceleration while ICE vehicles perform exceedingly well at constant speeds, the vehicles utilizing the gasoline ICE, electric motor, and battery can outperform the conventional vehicles as well as those pure EVs.

Besides changing the energy source, the entire vehicle structure of the automobile needs to be modified so as to solve the efficiency problem. Ever since the birth of the first modern automobile in 1885, the configuration of the vehicle steering mechanism has remained almost unchanged.* The configuration employed in early vehicles was similar to that used in the four-wheel automobiles developed afterward. The translation of the vehicle's center is highly dependent on the rotation of the vehicle attitude, and vice versa. Hence, it is difficult to achieve omnidirectional motion, i.e., the ability to translate in an arbitrary direction while the vehicle is oriented in another arbitrary direction.

To make a vehicle omnidirectional requires decoupling the translation motion from rotation. One way to measure the mutual dependency of translation and rotation is by the minimal turning radius. A minimal turning radius means the coupling of the two motions is small, which leads to the holy grail of omnidirectionality. The ideal minimal turning radius is zero, where no translation appears as a by-product of steering, which is most desirable. However, the front-wheel steering angle of a traditional vehicle is confined to the range from -35° to 35° because of the steering tie rod.[43] Such a rod hinders the vehicle from achieving a smaller minimal turning radius.

Engineers from the automative industry have endeavored to enhance the steerability of vehicles by introducing four-wheel steering (4WS) systems. The first 4WS vehicle appeared in 1938 and was manufactured by Mercedes-Benz. Its rear wheels were steered in the opposite direction to the front wheels to reduce the turning radius. On the other hand, in 1987, Honda

---

*Although the first vehicle had three wheels, the two rear wheels were conventional fixed wheels and the front wheel was a conventional center-oriented wheel.

started manufacturing the Prelude 4WS vehicle in which the steering angle of the rear wheels changes depending on the front wheels. When the steering wheel is turned slightly, the rear wheels will be steered in the same direction as the front wheels to improve stability during a high-speed turn or when changing lanes. When the steering wheel is turned at a large angle, the rear wheels will be steered in the opposite direction to that of the front wheels so as to shorten the turning radius. But still, the existence of the steering tie rod deters any further enhancement in steerability.

The recent release of Jeep Hurricane by the Chrysler Group has caught the attention of the automobile industry. The innovative vehicle is capable of zero radius turning, i.e., rotating about the center of the vehicle. It is four-wheel drive (4WD) with the power origins of the four wheels coupled with two engines. As a result, a complicated device is required to control the direction of each wheel since the directions of the four wheels depend on the driving mode, in either traditional front-wheel steering (FWS) or zero radius turning mode. Moreover, although its steering system is independent of translation, it does not allow for translation in arbitrary directions.

In parallel to the development from the automobile industry, some researchers attempt to borrow the concept from the field of robotics. Wheeled mobile robots capable of omnidirectional motions have been studied for more than a decade. Fujisawa et al. have built a platform based on four wheels with perpendicular rollers on the rims so as to enable parallel motion.[26] However, this is only a small-sized platform, and heavy loads for human carrier will crash the wheel mechanism. There are other platforms proposed in Refs. 59, 60, 89, 106, and 114, but they are not developed for the objective of functioning as human carriers.

An intelligent omnidirectional HEV integrating the latest robotics vehicle technology is being developed at the Chinese University of Hong Kong. Unlike the engine driving principle used in traditional vehicles, its engine is disconnected from the transmission and is connected to a generator that charges a battery package instead. Four in-wheel motors powered by the battery package are implemented to create the propelling force. This drive-by-wire technology therefore significantly reduces the complexity associated with mechanical differentials such as the one seen in the Jeep Hurricane. The steering tie rod is replaced by four sets of novel steering mechanisms which enable each wheel to be steered independently in the range of -35° to $+90^\circ$. Such a large range is enough not only for zero radius turning, but also for lateral parking in which translation is independent of steering. Thus, the problem of parking can be alleviated.

1.2 Objectives

The main objective of this book is to propose the novel intelligent omnidirectional HEV system to overcome the problems mentioned above. The proposed HEV consists of several distinct components, and each is a challenge to research. Unlike in the conventional vehicle, the mechanical and electrical system of the omnidirectional vehicle is fundamentally different. Apart from the vehicle structure (as some additional functions such as zero radius turning and lateral parking can be achieved with the four-wheel independency), new control mechanisms have to be devised including the four-wheel independent steering control and force control. In addition, for electric vehicles the battery is the most essential and critical component. Designing an efficient and robust battery management system is an extremely challenging task. In addition, intelligent energy management strategy is an important issue to manage the power flow of HEV and achieve balanced performances on fuel economy, emissions, grade ability, and acceleration ability. The following outlines the main contributions of this book:

1. Develop the omnidirectional structure, i.e., four-wheel independent steering (4WIS) structure, to release the dependence between translation and rotation motions by combining with the technology of robotics. It enables the vehicle to turn about any instantaneous center of rotation.
2. Develop an omnidirectional steer-by-wire system for 4WIS vehicles which is suitable for any 4WIS vehicles that need quick and high maneuverability.

3. Propose an optimal traction force distribution for 4WIS vehicles to improve vehicle stability without using any feedback information on vehicle motion.
4. Propose a deterministic yaw moment controller for 4WIS vehicles to enhance driving stability and controllability.
5. Propose some novel methodologies and solutions to overcome the challenging issues in the battery management system (BMS), such as state of charge (SOC) estimation, fast charging control, and cell equalization control.
6. Propose a power management system based on model predictive control (MPC) to optimize the efficiency of fuel consumption and performance of HEVs.

1.3 Outline of the Book

This book starts by discussing the hardware design of the 4WIS HEV. Then it gradually changes the focus to the software part which is responsible for the intelligence of the HEV. Here *intelligence* can refer to the optimization in traction force control, BMS, and energy management system as well. The remainder of this book is arranged as follows.

Chapter 2 elaborates on the system design and implementation of the independent and driving vehicle. The vehicle structure of 4WIS is implemented to realize omnidirectional motion.

Chapter 3 continues the discussion of the 4WIS vehicle and presents an omnidirectional steer-by-wire system. The system consists of an extended steering interface and a behavior-based steering controller. Its mathematical model and the concept and prototype of the extended steering interface are described. The design of the force feedback controller is presented. The behavior-based steering controller is introduced with the implementation and testing results discussed.

After discussing of the independent steering issue, the driving force for the wheels can be independent. This introduces the concept of four-wheel independent drive (4WID). In Chap. 4, we propose an optimal traction force distribution and a deterministic yaw moment controller for omnidirectional 4WIS and 4WID vehicles to enhance driving stability and controllability. It involves two sections. The first one presents the deterministic force generation algorithm and the optimal traction force distribution with the simulation results. The second section presents the direct yaw moment control that includes the wheel-slip observer, yaw rate error observer, and heading-error observer. Numerical simulation results are presented to evaluate its performance.

Chapters 2 to 4 complete the introduction of 4WIS vehicles. The subsequent chapters then start to focus on the electrical part of the HEV. The battery system of the EV, which is the most important but difficult aspect of EV design, hinders its mass commercialization. Hence, the BMS is discussed in Chap. 5 which is composed of four principal sections. Section 5.1 illustrates the system structure and hardware design of the BMS prototype. Section 5.2 focuses on the robust SOC estimation framework. Section 5.3 proposes a fast charge framework based on model predictive control (MPC), with the aim of simultaneously reducing charge duration, which represents safety and energy efficiency during the charging process. Section 5.4 introduces a SOC-based fuzzy controller to equalize cells. Both the simulations and experiments are presented to verify the efficacy of the proposed methods.

After introduction of the battery management system in Chap. 5, Chap. 6 describes how to combine it with the four-wheel independent steering vehicle proposed earlier to form an omnidirectional HEV. Chapter 6 proposes a new power management scheme for HEVs. Here we focus on the series HEV. First the model of series HEV is described in detail. Then a load forecasting algorithm is developed based on fuzzy logic and neural network classification. By fuzzy logic, the drive load is divided into five different levels. Next, by integrating the MPC and the load forecasting algorithm, a real-time power control system is introduced. Particle swarm optimization is used to find optimal operational weights for different drive levels. It adjusts operational weights automatically to achieve best performance. Finally, the implementation and experiments of the proposed method are presented to prove the proposed approach is practical and efficient.

Finally, Chap. 7 concludes the book and suggests some future research directions of intelligent omnidirectional HEVs.

CHAPTER 2

System Design and Implementation

In this chapter, the design and implementation of omnidirectional HEV are proposed. In Sec. 2.1, kinematic models of three modes are discussed, including the most general motion, i.e., rotation about an arbitrary instantaneous center of rotation (ICR), zero radius turning (ZRT), and two-front-wheel steering (2FWS). Section 2.2 presents the system design consisting of mechanical design, electric design, design for positioning of steering angles, software, and the user interface. Experiments on ZRT and lateral parking (LP) are illustrated in Sec. 2.3 for demonstration of omnidirectional motions of the proposed omnidirectional HEV.

2.1 Configuration

In most driving situations, especially when the vehicle is driven at high speed, 2FWS is sufficient for manipulation. For analysis of the kinematic and dynamic model of the vehicle, a single-track model was developed[92] which simplifies the four-wheel vehicle model.

However, when the vehicle is operated in a small and limited zone, steerability becomes significantly more important. Aiming for higher steerability, the proposed omnidirectional HEV breaks the invariant coupling relationship of pairs of front and rear wheels. It has three driving modes, namely, ZRT, LP, and 2FWS. In different modes, the coupling relationships of the wheels are different, but they all satisfy Ackerman's law.[92]

Figure 2.1 shows the kinematic model in which the vehicle turns about an arbitrary ICR, denoted as $[x_{\mathrm{ICR}_b}, y_{\mathrm{ICR}_b}]$.

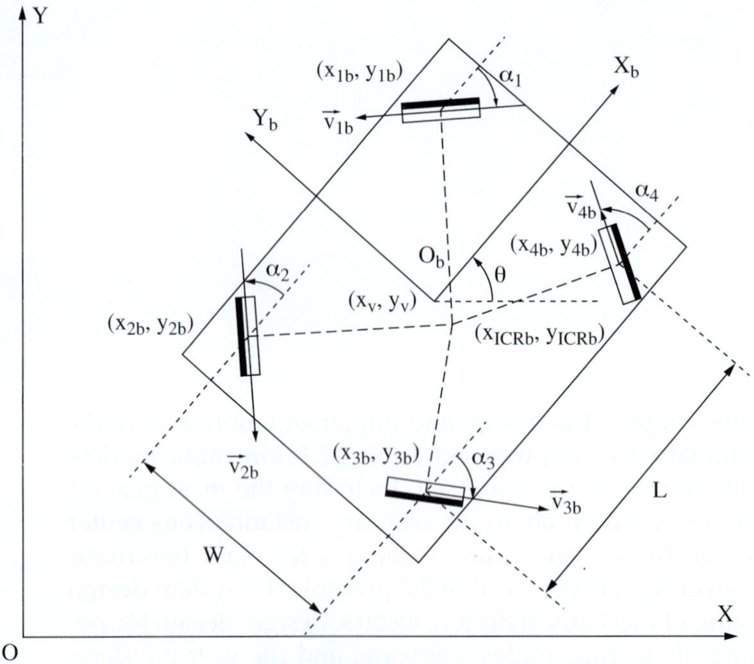

FIGURE 2.1 Kinematics of rotation about an arbitrary ICR.

OXY and $O_b X_b Y_b$ are the global coordinate frame and local coordinate frame fixed on the vehicle, respectively. The configuration of the vehicle can be expressed by $[x_v, y_v, \theta]$. Also $[x_{ib}, y_{ib}]$, $i = 1, 2, 3, 4$, is 2D coordinates of the centers of the four wheels in the local coordinate frame, with the sequence from left front, left rear, right rear, to right front. Each outer part of the wheel is marked black for convenience to identify its direction. And α_i, $i = 1, 2, 3, 4$, is the steering angle from the original position. All the angles are positive if the curve arrows are counter clockwise. And $\mathbf{v}_{ib}$, $i = 1, 2, 3, 4$, is the velocity vector in the local coordinate frame; W and L are wheel base and axis base, respectively.

By Ackerman's law, all the axes of the wheels must intersect at the ICR to permit motion. Thus the steering angles of the wheels are constrained by the equations below, which means that the line from each wheel center to the ICR should be perpendicular to the direction of wheel. This constraint reduces the steering degrees of freedom from 4 to 2.

$$[\cos(\alpha_i),\ \sin(\alpha_i)] \cdot ([x_{i_b},\ y_{i_b}]^T - [x_{\mathrm{ICR}_b},\ y_{\mathrm{ICR}_b}]^T) = 0 \qquad (2.1)$$

where $i = 1, 2, 3, 4$ and $[\cdot, \cdot]$ is a 2D row vector.

Moreover, because all four wheels have identical rotational velocity about the same ICR, equations of linear velocity clearly obey the equations below with the assumption of no friction.

$$\frac{([x_{\mathrm{ICR}b},\ y_{\mathrm{ICR}b}]^T - [x_{1b},\ y_{1b}]^T) \times \mathbf{v}_{1b}}{|[x_{\mathrm{ICR}b},\ y_{\mathrm{ICR}b}]^T - [x_{2b},\ y_{2b}]^T|^2}$$

$$= \frac{([x_{\mathrm{ICR}b},\ y_{\mathrm{ICR}b}]^T - [x_{2b},\ y_{2b}]^T) \times \mathbf{v}_{2b}}{|[x_{\mathrm{ICR}b},\ y_{\mathrm{ICR}b}]^T - [x_{2b},\ y_{2b}]^T|^2} \qquad (2.2)$$

$$= \frac{([x_{\mathrm{ICR}b},\ y_{\mathrm{ICR}b}]^T - [x_{3b},\ y_{3b}]^T) \times \mathbf{v}_{3b}}{|[x_{\mathrm{ICR}b},\ y_{\mathrm{ICR}b}]^T - [x_{3b},\ y_{3b}]^T|^2} \qquad (2.3)$$

$$= \frac{([x_{\mathrm{ICR}b},\ y_{\mathrm{ICR}b}]^T - [x_{4b},\ y_{4b}]^T) \times \mathbf{v}_{4b}}{|[x_{\mathrm{ICR}b},\ y_{\mathrm{ICR}b}]^T - [x_{4b},\ y_{4b}]^T|^2} \qquad (2.4)$$

There are three special cases in omnidirectional locomotion: ZRT (rotation is independent of translation), LP (translation is independent of rotation), and 2FWS (normal vehicle locomotion). Figure 2.2 illustrates their wheels' steering configuration in kinematics.

2.1.1 Zero Radius Turning

To change the attitude of the vehicle, traditional vehicles steer two front wheels. Position change of the center is a side effect of rotation. However, in narrow space, this side effect is not desired by drivers. Hence, it is ideal to rotate the vehicle without translation, which can be realized by ZRT.

In Fig. 2.2*a*, the right front and left rear wheels are steered to the same positive angle $\alpha = \arctan(L/W)$ and the other two are steered to the same negative angles $-\alpha$. In the ZRT mode, all the wheel speeds are numerically the same.

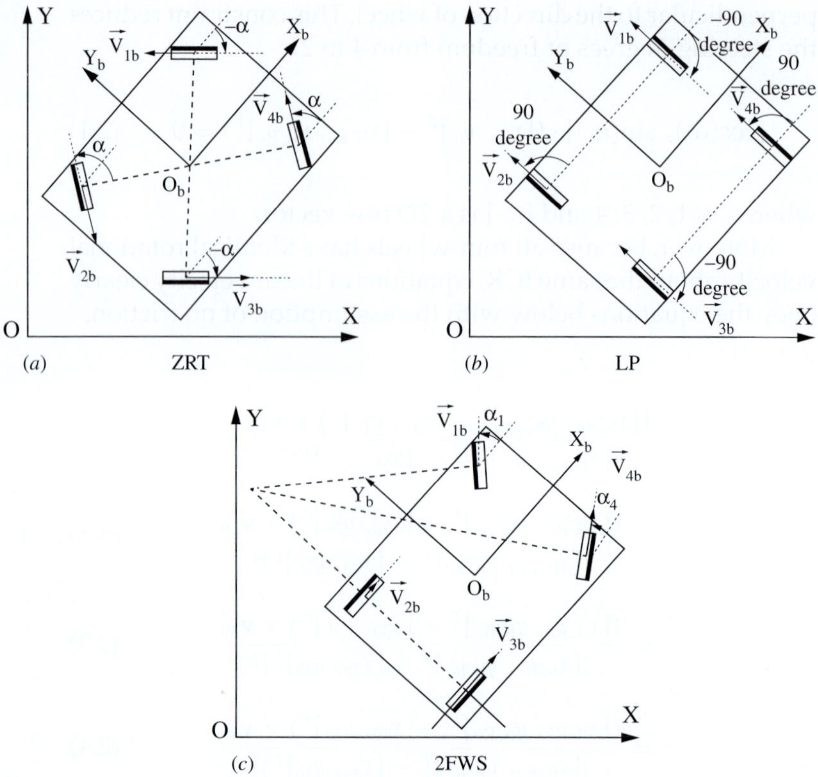

FIGURE 2.2 Steering configurations of wheels.

For a 360° change in attitude, the turning radius and the occupied space for ZRT, 2FWS, and 4WS by derivation are compared as follows.

ZRT:

$$R_{\mathrm{ZRT}} = 0 \tag{2.5}$$

$$\mathrm{Area}_{\mathrm{ZRT}} = \frac{\pi(W^2 + L^2)}{4} \tag{2.6}$$

2FWS:

$$R_{\mathrm{2FWS}} = \sqrt{\left(Lctg\alpha_{\max} + \frac{W}{2}\right)^2 + \frac{L^2}{4}} \tag{2.7}$$

	Turning Radius (m)	Occupied Area (m^2)
ZRT	0	3.76
2FWS	3.91	33.97
4WS	1.90	16.22

TABLE 2.1 Comparison of Space Requirement

$$\text{Area}_{2\text{FWS}} = \pi(2LWctg(\alpha_{\max}) + W^2 + L^2) \quad (2.8)$$

4WS:

$$R_{4\text{WS}} = \frac{L}{2}ctg\alpha_{\max} + \frac{W}{2} \quad (2.9)$$

$$\text{Area}_{4\text{WS}} = \pi\left(LWctg\alpha_{\max} + W^2 + \frac{L^2}{4}\right) \quad (2.10)$$

Table 2.1 compares occupied space among ZRT, 2FWS, and 4WS. Note that the most space-saving motion is ZRT.

2.1.2 Lateral Parking

Besides rotation, translation can also be manipulated directly and independently. For pure translation, i.e., motion without change in attitude, all four wheels ought to be steered to an identical angle. This can be represented by the Ackerman model that the ICR intersects at a point at infinity. For simplification in vehicle manipulation, LP is the most desirable function because in most parking situations, the parking slot is only 1 or 2 m parallel to the current vehicle position.

In Fig. 2.2*b*, the left front and right rear wheels are steered to $-90°$, and the other two are steered to $+90°$. The four velocity vectors are parallel and have the same length under the skid-free assumption.

2.1.3 Front-Wheel Steering

As the traditional 2FWS is still popular and in general cases an applicable means of motion, it is worthwhile to preserve it for manipulation.

In Fig. 2.2*c*, the ICR cannot be arbitrary but must be on the line where the axes of the two rear wheels are collinear.

In this mode, we take into consideration the mapping of the angle input of the steering wheel to the angles of the two front wheels. For simplicity, the angle mapping is piecewise. In any turning manipulation, the angle of the inner front wheel is directly proportional to the angle of steering wheel input, $\alpha_{\text{innerfront}} = K\phi_{\text{steeringwheel}}$. Considering the 35° constraint in the traditional vehicle, it is clear that when $\phi_{\text{steeringwheel}}$ is turned to the maximum $\alpha_{\text{innerfront}} = 35°$. Hence the gain K can be identified.

The steering angle of the outer front wheel shall obey the Ackerman law, and thus $\alpha_{\text{outerfront}} = \arctan(\frac{L}{W+Lctg\alpha})$.

2.2 System Design

The design of the omnidirectional HEV is novel to traditional vehicles. It is a drive-by-wire and steer-by-wire system. For the functionality of omnidirectional motion, four-wheel independent driving (4WID) and four-wheel independent steering (4WIS) are necessary for the system. Figure 2.3 shows the look of the vehicle, and Table 2.2 illustrates its specifications.

FIGURE 2.3 Look of the omnidirectional vehicle.

Tread	1.1 m
Wheelbase	1.9 m
Mass	300 kg
Driving motor	48 V, 600 W
Steering motor	42 V, 90 W
Battery	48 V, 44 Ah

TABLE 2.2 System Specification of Omnidirectional HEV

2.2.1 Mechanical Design

The kernel architecture of an omnidirectional HEV consists of a chassis and four independent steering mechanisms, as illustrated in Fig. 2.4.

In the steering operation, the group including an in-wheel motor, its outer frame, a disc brake, an upper and a lower arm, and two spring dampers is steered about the rotational shaft, which in the upper end is driven by the output shaft of the speed reduction gearbox. The input of the gearbox is a dc motor. The in-wheel motor, which is for driving, is originally a brushless dc motor for the electric motorcycle, which is controlled by the voltage signal from the acceleration pedal. As the omnidirectional HEV is a platform for an on-road vehicle, for an absorbing vertical shock the suspension system is preserved. Totally, there are four such independent steering mechanisms, one for each driving wheel.

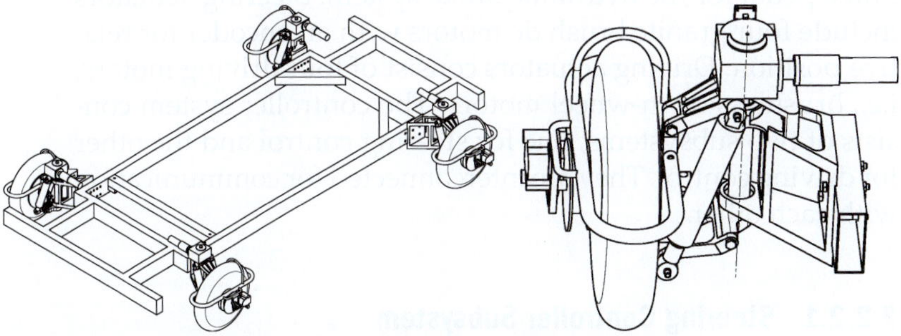

(*a*) (*b*)

FIGURE 2.4 (*a*) Chassis and (*b*) steering mechanism.

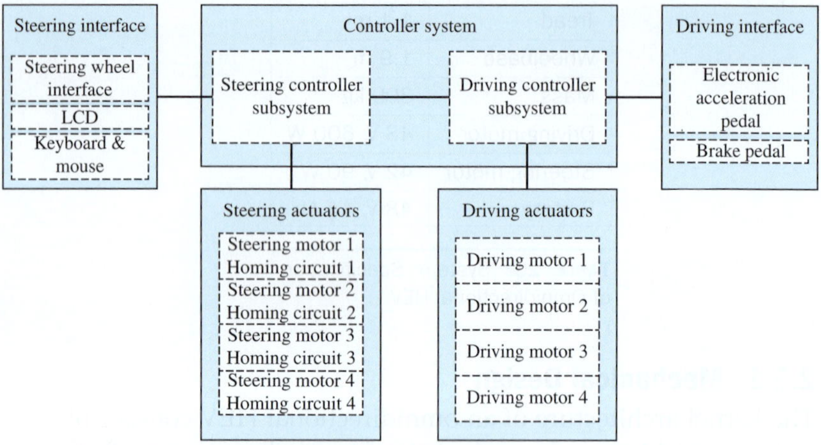

FIGURE 2.5 Controller system.

2.2.2 Electric Design

Figure 2.5 is the outline of the electric design of an omnidirectional HEV. There are five parts in the system: the controller system, steering interface, driving interface, steering actuators, and driving actuators.

Inside the steering interface, there is a racing steering wheel, which has a shift bar and six buttons for direction and mode selection. An LCD, a keyboard, and a mouse are assembled for the purpose of debugging and testing. The driving interface is composed of two pedals, an electronic acceleration pedal and a brake pedal for the hydraulic brake system. Steering actuators include four granite brush dc motors with an encoder for relative position. Driving actuators consist of four driving motors, i.e., brushless dc in-wheel motors. The controller system consists of two subsystems, one for steering control and the other for driving control. They are interconnected for communication with each other.

2.2.2.1 Steering Controller Subsystem

The steering controller subsystem is illustrated in Fig. 2.6. A PC104 with the configuration of an onboard Intel Pentium

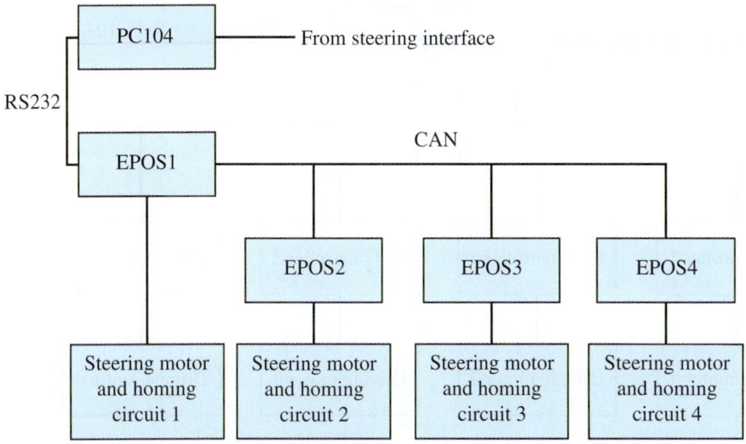

FIGURE 2.6 Steering controller subsystem.

M1.1-GHz, 1-GB DDR RAM, and a 1-GB CF card as hard disk is the kernel controller in the steering controller subsystem. Linux DSL 2.4 is adopted as the OS for the PC104.

The PC104 sends commands directly to steering motor controllers, and each command is labeled with an ID. Maxon EPOS is utilized for position control of the steering motor. It is equipped with either CAN bus or RS232 for communication. The omnidirectional HEV utilizes the configuration of RS232-CAN gate connection.

2.2.2.2 Driving Controller Subsystem

In the driving controller subsystem (Fig. 2.7), there is a DSP (TMS320F2812) as the kernel controller and four brushless dc motor controllers.

The DSP obtains the voltage signal from the electronic acceleration pedal and sends it to the four driving motor controllers. Complexity in the differential part of the traditional vehicle has been greatly reduced in this drive-by-wire subsystem. The directions of the four driving motors are different for each of the six patterns (three modes and each with two directions), as shown in Table 2.3, where B and F denote backward and forward, respectively.

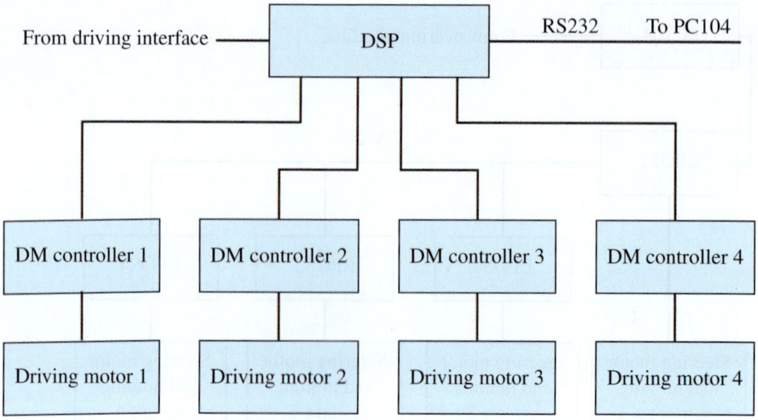

FIGURE 2.7 Driving controller subsystem.

2.2.3 Positioning of Steering Angles

In the design of the steering subsystem, the absolute position of the steering angle for each steering mechanism is a significant issue, because the encoder in the steering motor is only effective as an relative position sensor, and calibration will be required for each power-on step. Research on the calibration of steering angles has been carried out by McKerrow and Ratner.[62] However, this complicated calibration method has to be conducted for every power-on step.

In the omnidirectional vehicle, we utilize an optical sensor as a homing sensor for each steering motor and relieve the trouble of calibration. Figure 2.8 illustrates this concept. A blocking

	Forward (♮1 ♮2 ♮3 ♮4)	**Backward (♮1 ♮2 ♮3 ♮4)**
2FWS	(B B F F)	(F F B B)
ZRT	(F F F F)	(B B B B)
LP	(F B B F)	(B F F B)

TABLE 2.3 Directions of Driving Motors

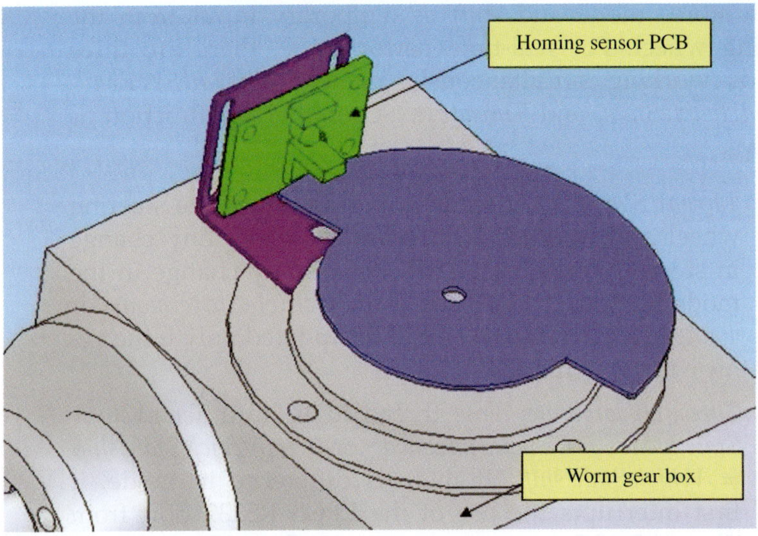

FIGURE 2.8 Homing device.

piece, with the shape at a semiring edge cut off a circle plate, is fixed to the upper end of the output shaft of the speed reduction gearbox. The edge of the voltage change from the optical sensor is an ideal means for absolute position calibration.

2.2.4 Software

Because of the large quantity of controllers and ports, i.e., two kernel controllers, eight peripheral controllers, two RS232 ports, and one USB port, much attention has to be paid to the synchronization to prevent conflict. Within the two kernel controllers, the PC104 is a master while the DSP works as a slave.

2.2.4.1 PC104 Program

In the PC104 algorithm, there are two structure variables. One is called *VehicleState* for the storage of mode, shift, angle of the steering wheel, and velocity. The other is *TempVehicleState*, which stores mode and shift as temporary input from the steering wheel. For real-time performance, there are three threads working simultaneously, namely, *ThreadSWRead*, *ThreadUpdateState*, and *ThreadEPOSControl*. They function as follows:

1. *ThreadSWRead*. It reads a message from the steering wheel and updates *VehicleState.angle* if any change in steering angle occurs. If there is any change in the mode button or shift bar, *TempVehicleState.mode* or *TempVehicleState.shift* will be updated only if the vehicle is not in motion.
2. *ThreadUpdateState*. This thread idles until *TempVehicleState.mode* $\neq$ *VehicleState.mode* or *TempVehicleState.shift* $\neq$ *VehicleState.shift*. If there is difference in mode, it first interrupts the use of the EPOS RS232 port from *ThreadEPOSControl* to prevent conflict. Thereafter, it sends a command to each EPOS to initialize to the new mode. With success, it then starts communication with DSP to change its mode state. It finally sets the new mode to *VehicleState.mode*. If there is difference in shift, it transmits that information to the DSP, and with a successful acknowledgment it then updates *VehicleState.shift*.
3. *ThreadEPOSControl*. In this thread, the algorithm calculates the four steering angles according to *VehicleState.mode*, and sets those to the target angles for the four EPOS controllers. But if there is any signal from *ThreadUpdateState* to release the possession of the EPOS RS232 port, it will pause communication with EPOS and start idling.

2.2.4.2 DSP Program

DSP supports single thread, and the program has the following functions:

1. To obtain the vehicle velocity. The velocity is sent to PC104 via the RS232 port.
2. To digitize the voltage signal of the electronic acceleration pedal, calculate the acceleration voltage signals for four driving motor controllers and output them.

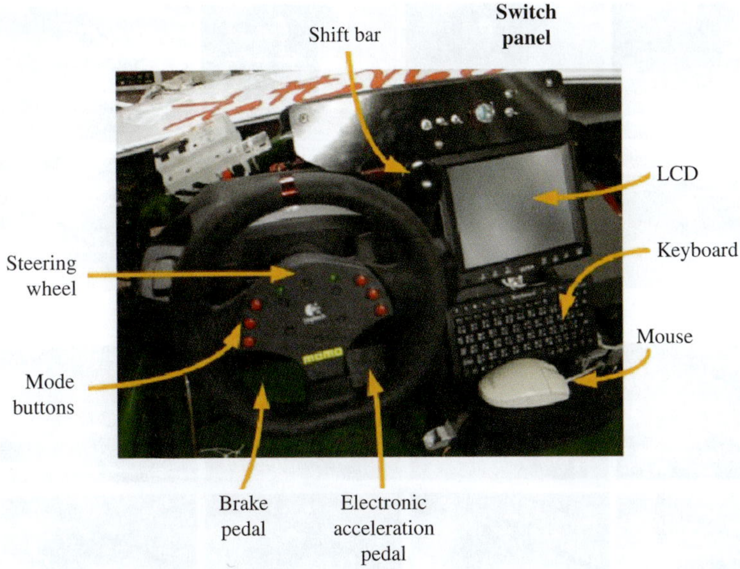

FIGURE 2.9 User interface.

3. In case of any mode/shift change demand from the PC104, it will change the mode/shift state in DSP and reply if successful.

2.2.5 User Interface

Figure 2.9 illustrates the user interface. We adopt Logitech Momo as the steering wheel and the buttons on it for mode change to "Idle Operation," "2FWS," "ZRT," and "LP." Shift bar is integrated into the steering wheel device as well.

The electronic acceleration pedal and braking pedal are located in the same place as in traditional vehicles.

2.3 Implementation and Testing

With the developed platform of the omnidirectional HEV, we realize the functionality of ZRT and LP and demonstrate its significant advantage in parking scenes. The test (Fig. 2.10) is

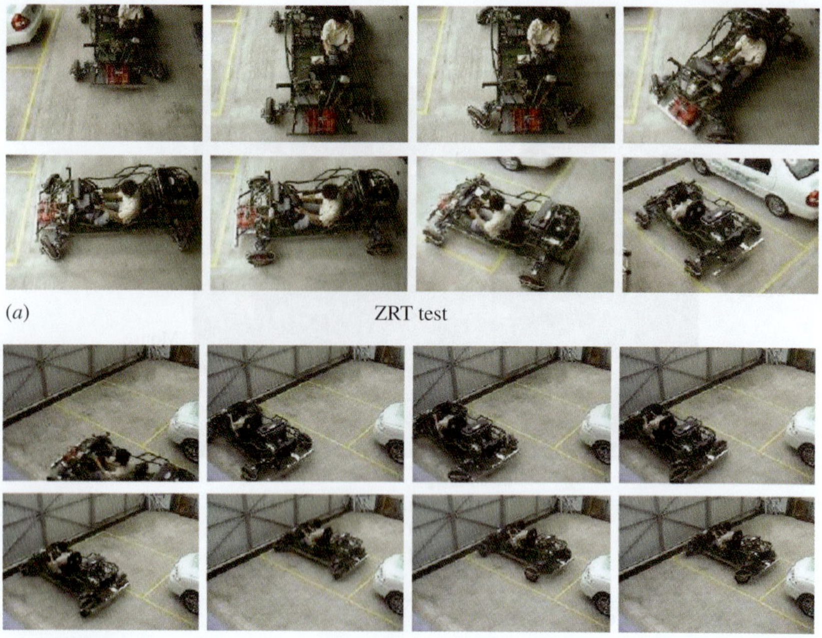

(a) ZRT test

(b) LP test

FIGURE 2.10 Experiment on parking in two different slot arrays.

carried out in scenes with two most popular parking slot arrangements, i.e., parallel and serial.

In the experiment, the omnidirectional HEV performs well with both functions of ZRT and LP. The demand on the driver for trajectory selection and generation is greatly relieved. The decoupling of rotation and translation enables the driver to manipulate the vehicle more freely and easily. However, because the adoption of the speed reduction gearbox inevitably brings gaps into the steering mechanism, i.e., the gap between the input gear and output gear, each wheel has an error of about $\pm 3^\circ$. Consequently, 2FWS and ZRT are demonstrated as satisfactory in the test, while translation in LP meets a bit of a challenge due to the increased lateral friction caused by the gear gap.

2.4 Summary

This chapter presents the system and design of an omnidirectional HEV with the functionality of ZRT and LP as well as 2FWS, similar to the conventional vehicle. Benefiting from the features of 4WIS and 4WID, the HEV is a platform that can be incorporated with other locomotion, such as linear translation, i.e., translation without change in steering directions, etc.

CHAPTER 3

Four-Wheel Independent Steering Control

This chapter introduces the four-wheel independent steering (4WIS) control. Section 3.1 describes the mathematical model of the 4WIS vehicle. In Sec. 3.2, the concept and prototype of the extended steering interface are presented. Section 3.3 presents the design of the force feedback controller. Furthermore, the behavior-based steering controller is introduced in Sec. 3.4. The architecture of the HIL simulation and the simulation results are discussed in Sec. 3.5.

3.1 Modeling

This section describes the mathematical model of the 4WIS vehicle. The notations and coordinate system of a vehicle model are shown in Fig. 3.1. The numbers 1, 2, 3, and 4 represent left front, left rear, right rear, and right front wheel, respectively. The coordinates of the oriented center of wheels are denoted as (x_i, y_i), where $i = 1, 2, 3, 4$. The orientations of the wheels can be adjusted about the center of wheel by the steering actuator, and it is denoted as δ_i. And v_x and v_y are the velocities of the center of mass in the x and y direction, respectively, while ω denotes the angular velocity of the vehicle body. To evaluate the performance of the proposed controller, the dynamic model of the vehicle body, tire, and steering actuator is adopted in the numerical simulations.

3.1.1 Vehicle Body

The vehicle model only considers the longitudinal, lateral, and yaw dynamics. Vertical, roll, and pitch dynamics are neglected.

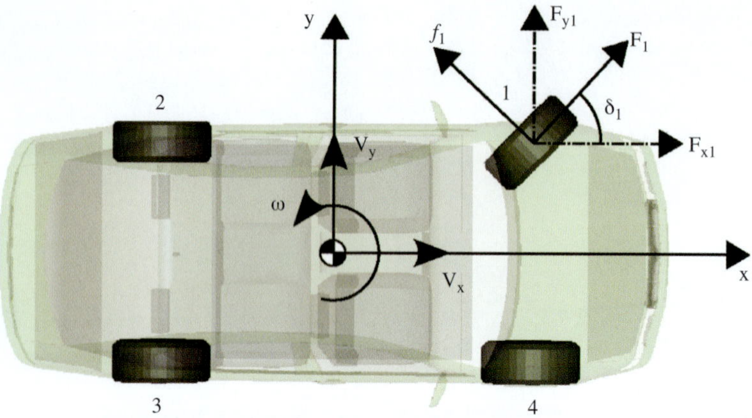

FIGURE 3.1 Notations and coordinate system of vehicle.

The weight of vehicle is assumed to be equally distributed on each wheel. Hence, the vertical tire force on each wheel is $mgN/4$. According to Ref. 92, the dynamic equations of the vehicle body can be written as follows:

$$\sum_{i=1}^{4} F_{x_i} = m\left(\dot{v}_x - v_y\omega\right) \tag{3.1}$$

$$\sum_{i=1}^{4} F_{y_i} = m\left(\dot{v}_y + v_x\omega\right) \tag{3.2}$$

$$\sum_{i=1}^{4} \left(F_{y_i}x_i - F_{x_i}y_i\right) = I_z\dot{\omega} \tag{3.3}$$

where m and I_z denote the mass and yaw moment of inertia of vehicle, respectively; F_{x_i} and F_{y_i} are the x- and y-component forces applied on the steering joints. It can be determined by

$$F_{x_i} = F_i \cos\delta_i - f_i \sin\delta_i \tag{3.4}$$

$$F_{y_i} = F_i \sin\delta_i + f_i \cos\delta_i \tag{3.5}$$

3.1.2 Tire

Wheel generates force on the vehicle body through the steering joint. The longitudinal force on wheel F_i is generated by the driving actuator while the lateral force on wheel f_i is generated by wheel slip in the lateral direction. The longitudinal force on the wheel is assumed to be directly proportional to the torque generated by the driving actuator while the lateral force on the wheel can be approximated by the Magic Formula Tire Model proposed by Pacejka[2]

$$f(\alpha_i) = D\sin(C\arctan\{B\alpha_i - E[B\alpha_i - \arctan(B\alpha_i)]\}) \quad (3.6)$$

where B, C, D, and E are the tire parameters. Here α_i represents the angle of slip of wheel i, and it can be determined by

$$\alpha_i = \delta_i - \operatorname{atan2}\left(v_y + \omega x_i,\ v_x - \omega y_i\right) \quad (3.7)$$

3.1.3 Steering Actuator

The steering modules on each wheel are identical and driven independently. The steering axis is perpendicular to the ground and passes through the center of the wheel. DC motors are used as the steering actuators. The motor model can be described by the following equations:

$$L\frac{di}{dt} = V - Ri - K_b\omega_m \quad (3.8)$$

$$J_m\frac{d\omega_m}{dt} = K_m i - K_f\omega_m - T_m \quad (3.9)$$

where the description of the notation used can be found in Table 3.1. The motor should be equipped with a gear reducer so as to increase the output torque. The gear ratio must be chosen properly so that the output torque is able to resist road disturbance and meanwhile have reasonable turning speed. Assume there is no energy loss in the transmission; the dynamic relationships between motor and load are shown as follows:

$$\frac{T_m}{T_L} = \frac{\omega_L}{\omega_m} = n \quad (3.10)$$

$$J_L\frac{d\omega_L}{dt} = T_L + T_d \quad (3.11)$$

	Symbol	Unit	Value
Acceleration of gravity	g	$kg \cdot m^{-2}$	9.8
Vehicle mass	m	kg	1000
Vehicle moment of inertia	I_z	$kg \cdot m^2$	1600
Coordinates of wheel 1	(x_1,y_1)	m	(1.25,0.75)
Coordinates of wheel 2	(x_2,y_2)	m	(−1.25,0.75)
Coordinates of wheel 3	(x_3,y_3)	m	(−1.25,−0.75)
Coordinates of wheel 4	(x_4,y_4)	m	(1.25, −0.75)
Cornering coefficient	C_α	N/rad	25,000
Torque at output shaft	T_L	$N \cdot m$	—
Torque at motor shaft	T_m	$N \cdot m$	—
Disturbance on load	T_d	$N \cdot m$	—
Angular velocity of output shaft	ω_L	rad/s	—
Angular velocity of motor rotor	ω_m	rad/s	—
Output shaft load inertia	J_L	$kg \cdot m^2$	0.08
Motor rotor inertia	J_m	$kg \cdot m^2$	6.96×10^{-6}
Gear ratio	n	—	1/200
Source voltage	V	V	—
Current of the circuit	i	A	—
Motor terminal resistance	R	Ω	2.07
Motor terminal inductance	L	H	6.2×10^{-4}
Back-emf constant	K_b	$V \cdot s/rad$	0.0525
Armature constant	K_m	$N \cdot m/A$	0.053
Viscous friction	K_f	N.m.s/rad	1.4×10^{-3}
Force feedback gain (ϕ1)	$K_{\phi 1}$	—	10
Force feedback gain (ϕ2)	$K_{\phi 2}$	—	10
P gain for P control	K_p^p	—	10
I gain for P control	K_i^p	—	0.01
D gain for P control	K_d^p	—	1
P gain for K control	K_p^k	—	100
I gain for K control	K_i^k	—	10
D gain for K control	K_d^k	—	10

TABLE 3.1 The Notations and Values Used in Modeling and Simulations

Substitute (3.10) and (3.11) into (3.8) and (3.9), and the motor model with gear reducer becomes

$$L\frac{di}{dt} = V - Ri - nK_b\omega_L \tag{3.12}$$

$$\left(J_m + n^2 J_L\right)\frac{d\omega_L}{dt} = nK_m i - K_f\omega_L + n^2 T_d \tag{3.13}$$

3.2 Extended Steering Interface

The conventional steering interface only provides one steering input (rotation of steering wheel) that is not enough to define all ICR in the horizontal plane. As a result, extension of the steering interface is needed for the omnidirectional 4WIS vehicle. As the steer-by-wire technology is applied, the extended steering interface is mainly aimed at defining the target ICR and passes it to a steering controller. The design of the extended steering interface provides a user-friendly interface to locate ICR without doing any mode-switching action.

3.2.1 Interface Design

The concept of the proposed steering interface is shown in Fig. 3.2. It can be treated as an extension of a conventional steering interface. A conventional steering wheel is kept. It

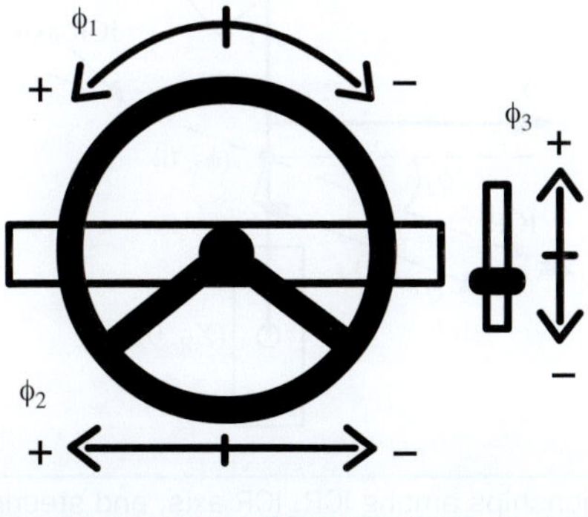

FIGURE 3.2 Extended steering interface.

is attached on a sliding rail so that the steering wheel has 2 degrees of freedom, i.e., rotation and translation. In addition, a sliding button is added beside the steering wheel; ϕ1, ϕ2, and ϕ3 represent the steering inputs. Force feedback actuators are connected to the rotation and translation joint on the steering wheel. All ICRs can be defined by the combinations of steering inputs. The handling of the extended steering interface is intuitive. The conventional steering behavior is reserved. Turning the steering wheel corresponds to turning motion. Additionally, sliding the steering wheel corresponds to pure translation motion. The joints can also be controlled simultaneously to represent the combined motion.

3.2.2 Defining Target ICR

Figure 3.3 illustrates the relationship among the ICR, ICR axis, and steering inputs. An abbreviated single-track model is employed to illustrate the concept. In the figure, $(x_F, 0)$, $(x_R, 0)$, and (ϕ3, 0) are the coordinates of the oriented center of front wheel, rear wheel, and origin of ICR axis (intersection point between ICR axis and x axis), respectively. The ICR always lies on the ICR axis. The origin of the coordinate system is located at the geometric center of the vehicle.

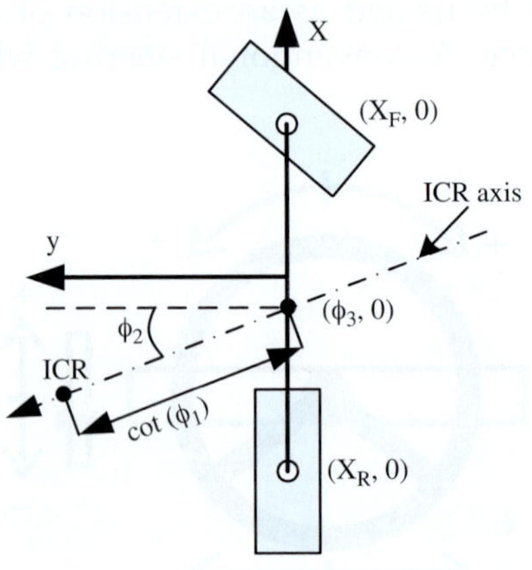

FIGURE 3.3 Relationships among ICR, ICR axis, and steering inputs.

Now cot ($\phi 1$) defines the distance between the ICR and the origin of the ICR axis; cot ($\phi 1$) > 0 represents left turning motion (turn anticlockwise about ICR) while cot($\phi 1$) < 0 represents right turning motion (turn clockwise about ICR). The steering result is equivalent to a conventional 2WFS system when $\phi 2 = 0$ and $\phi 3 = x_R$. To make the handling more intuitive, the left turning motion will be performed only when the steering wheel is turned left, and the right turning motion will be performed only when the steering wheel is turned right. As a result, the value of $\phi 1$ is restricted in $-\pi/2 \leq \phi 1 \leq \pi/2$. This restriction induces the limitation of the continuous motion of the ICR. The motion of ICR cannot pass through the origin of ICR.

Note that $\phi 2$ defines the angle between the ICR axis and the y axis. When $\phi 1 = 0$, changing the value of $\phi 2$ makes the vehicle performs pure translation in a different direction (Fig. 3.4*d*). To make the handling more intuitive, the left translation motion will be performed only when steering wheel is slid left, and the right translation motion will be performed only when the steering wheel is slid right. As a result, the value of $\phi 2$ is restricted in $-\pi \leq \phi 2 \leq \pi$.

Now $\phi 3$ defines the position of the origin of the ICR axis. Theoretically, using $\phi 1$ and $\phi 2$ is already enough to define the location of ICR in the 2D plane. The additional steering input $\phi 3$ is introduced to allow emulation of several kinds of conventional steering behaviors that make steering more intuitive. For instance, when $\phi 2 = 0$ and $\phi 3 = x_R$, the vehicle is

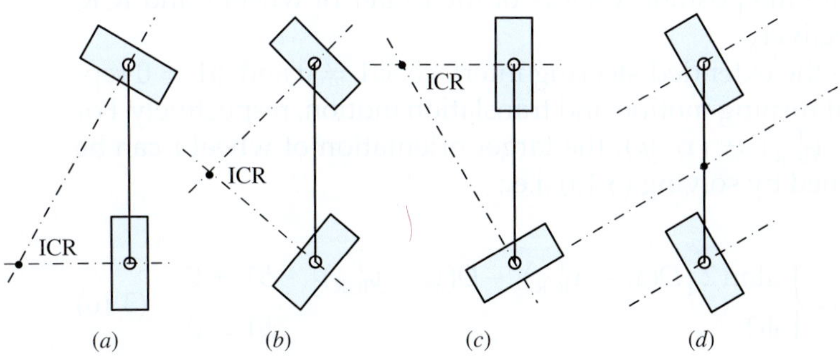

FIGURE 3.4 Examples of steering results.

performing typical front two-wheel steering (2WS) (Fig. 3.4*a*); when $\phi 3 = (x_F + x_R)/2$, the vehicle is performing typical 4WS (Fig. 3.4*b*); when $\phi 3 = x_F$, the vehicle is performing typical rear 2WS (Fig. 3.4*c*).

The coordinates of target ICR $(x^T_{\text{ICR}}, y^T_{\text{ICR}})$ can be defined by following equation:

$$\begin{bmatrix} x^T_{\text{ICR}} \\ y^T_{\text{ICR}} \end{bmatrix} = \begin{bmatrix} \cos\phi 2 & -\sin\phi 2 \\ \sin\phi 2 & \cos\phi 2 \end{bmatrix} \begin{bmatrix} 0 \\ \cot\phi 1 \end{bmatrix} + \begin{bmatrix} \phi 3 \\ 0 \end{bmatrix} = \begin{bmatrix} \phi 3 - \cot\phi 1 . \sin\phi 2 \\ \cot\phi 1 . \cos\phi 2 \end{bmatrix} \tag{3.14}$$

With the combination of these steering inputs, the vehicle can be commanded to turn about any ICR or translate in any direction. No mode-switching operation is needed.

3.2.3 Defining Target Orientations of Wheels

Once the target ICR is defined, the target orientations of the wheels can be determined. Under the kinematic constraint, the orientations of wheels must comply with

$$\omega \times (\mathbf{r}_i - \mathbf{r}_{\text{ICR}}) = \mathbf{v}_i \tag{3.15}$$

where $\mathbf{v}_i$ is the longitudinal velocity of wheel i and $\mathbf{r}_i$ and $\mathbf{r}_{\text{ICR}}$ are the position vectors of the center of wheel i and ICR, respectively.

In the extended steering interface, $\phi 1 \neq 0$ and $\phi 1 = 0$ represent turning motion and translation motion, respectively. For $(x^T_{\text{ICR}}, y^T_{\text{ICR}}) \neq (x_i, y_i)$, the target orientation of wheel i can be obtained by solving (3.15), i.e.:

$$\delta_i^T = \begin{cases} \operatorname{atan2}\left[D(x_i - x^T_{\text{ICR}}), -D(y_i - y^T_{\text{ICR}})\right] & \phi 1 \neq 0 \\ \phi 2 & \phi 1 = 0 \end{cases} \tag{3.16}$$

where $D = \text{sign}(\phi 1)$. The value of D will either be 1 or -1 to represent anticlockwise or clockwise turning, respectively.

When $(x_{\mathrm{ICR}}^{T}, y_{\mathrm{ICR}}^{T}) = (x_i, y_i)$, the target orientation of wheel i is undefined. The coordinates (x_i, y_i) are the singular points of steering. The wheel i does not roll but only twists on the ground when the vehicle turns. The tire will be worn. Besides, if the motion of ICR is passing through the singular point, the wheel has to be reoriented instantaneously by π rad. Even if the motion of ICR is not passing through the singular point but only close to it, it may also lead to a huge change of the target orientation. The steering actuator may not be fast enough to go to the target orientation. As a result, the target ICR is better kept away from the singular points.

3.2.4 Hardware Prototype

Figure 3.5 shows the hardware prototype of the extended steering interface. In the extended steering interface, the turning ($\phi 1$) and sliding ($\phi 2$) joints are connected with force feedback motors. The steering wheel is directly mounted on the output shaft of the force feedback motor while the slide rail is connected to the force feedback motor through the rack and pinion. Each motor is equipped with an optical encoder to obtain the posture of each joint. A microcontroller collects the posture of the

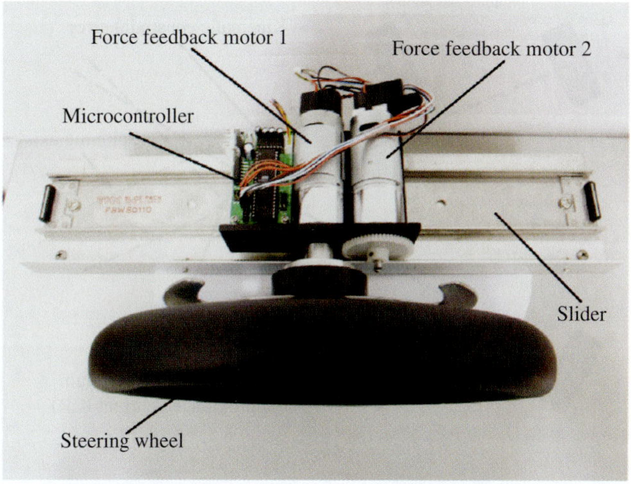

FIGURE 3.5 Prototype of the extended steering interface.

steering interface and sends it to the vehicle control system. The microcontroller also acted as the force feedback controller. It is accomplished by acquiring the appropriate interface posture from the vehicle control system.

3.3 Force Feedback Control

The actual steering condition should feed back to the steering interface so as to synchronize the target and actual steering motion. It helps the driver to realize the condition of the wheels and the ground so as to improve vehicle handling. The actual steering condition can be obtained by the current orientations of the wheels δ_i^C.

3.3.1 Defining Current Steering Condition

In turning motion, all the spinning axes of the wheels should intersect at one point, i.e., the ICR. However, this cannot always be achieved in the transient state due to road disturbance. Figure 3.6 illustrates the common situation of the orientations of

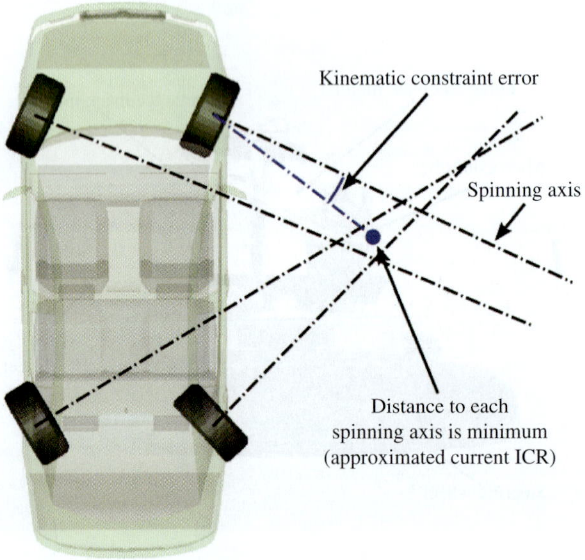

FIGURE 3.6 Approximated current ICR.

wheels. As a result, the current ICR $(x^C_{\text{ICR}}, y^C_{\text{ICR}})$ is approximated as the coordinates where the shortest distance to each spinning axis is minimum. The least-squares method is deployed to define the approximated current ICR, i.e., to find a set of coordinates where the sum of the squared shortest distances between the coordinates and the spinning axis of wheel i (d_i) is minimum:

$$\text{Minimize} \sum_{i=1}^{4} d_i^2 \tag{3.17}$$

The shortest distance d_i can be determined as follows:

$$d_i^2 = \left(y^C_{\text{ICR}} - m_i x^C_{\text{ICR}} - c_i\right)^2 \Big/ \left(m_i^2 + 1\right) \tag{3.18}$$

where $m_i = \tan\left(\delta_i^C + \pi/2\right)$ is the slope and $c_i = y_i - x_i m_i$ is the y intercept of the spinning axis of wheel i.

Taking partial derivatives of (3.17) with respect to x^C_{ICR} and y^C_{ICR}, we obtain

$$\begin{aligned} \sum_{i=1}^{4} \frac{\partial d_i^2}{\partial x^C_{\text{ICR}}} &= \sum_{i=1}^{4} \frac{2m_i^2 x^C_{\text{ICR}} - 2m_i y^C_{\text{ICR}} + 2c_i m_i}{m_i^2 + 1} = 0 \\ \sum_{i=1}^{4} \frac{\partial d_i^2}{\partial y^C_{\text{ICR}}} &= \sum_{i=1}^{4} \frac{2y^C_{\text{ICR}} - 2m_i x^C_{\text{ICR}} - 2c_i}{m_i^2 + 1} = 0 \end{aligned} \tag{3.19}$$

By solving (3.19), the approximated current ICR can be determined:

$$\left(x^C_{\text{ICR}}, y^C_{\text{ICR}}\right) = \left(\frac{BF - CE}{AE - B^2}, \frac{AF - BC}{AE - B^2}\right) \tag{3.20}$$

where $A = \sum_{i=1}^{4} \frac{2m_i^2}{m_i^2+1}$, $B = \sum_{i=1}^{4} \frac{2m_i}{m_i^2+1}$, $C = \sum_{i=1}^{4} \frac{2c_i m_i}{m_i^2+1}$, $E = \sum_{i=1}^{4} \frac{2}{m_i^2+1}$, and $F = \sum_{i=1}^{4} \frac{2c_i}{m_i^2+1}$.

For $AE - B^2 \neq 0$, the coordinates of current ICR can be found, and the vehicle is performing turning motion. When $AE - B^2 = 0$, the current ICR cannot be found. It implies all the spinning axes are in parallel. In this case, the vehicle is performing pure translation motion.

3.3.2 Force Feedback Reference

The force feedback controller uses $\hat{\phi}1$ and $\hat{\phi}2$ as references to feedback force to both steering joints $\phi1$ and $\phi2$. And $\hat{\phi}1$ and $\hat{\phi}2$ represent appropriate values of steering inputs $\phi1$ and $\phi2$ such that the target ICR is equal to the approximated current ICR.

In pure translation motion ($AE - B^2 = 0$), the values of $\hat{\phi}1$ and $\hat{\phi}2$ are defined as follows:

$$\hat{\phi}1 = 0 \tag{3.21}$$

$$\hat{\phi}2 = (\delta_1^C + \delta_2^C + \delta_3^C + \delta_4^C)/4 \tag{3.22}$$

In turning motion ($AE - B^2 \neq 0$), the values of $\hat{\phi}1$ and $\hat{\phi}2$ can be obtained as follows:

$$\hat{\phi}1 = D^C \cot^{-1}\left(\sqrt{(x_{ICR}^C - \phi3)^2 + y_{ICR}^{C\,2}}\right) \tag{3.23}$$

$$\hat{\phi}2 = \operatorname{atan2}\left(D^C(\phi3 - x_{ICR}^C),\, D^C y_{ICR}^C\right) \tag{3.24}$$

where D^C denotes the current turning direction. It can be found by determining whether the orientation of the wheel relative to $\mathbf{r}_i - \mathbf{r}_{ICR}$ is positive or negative:

$$D^C = \operatorname{sign}\left[\delta_i^C - \operatorname{atan2}\left(y_i - y_{ICR}^C,\, x_i - x_{ICR}^C\right)\right] \tag{3.25}$$

where i can be chosen arbitrarily except that the wheel that the oriented center is equal to the current ICR. As a result, it is suggested to choose the wheel for which the oriented center is farthest from the current ICR.

A little change of the current ICR that closes to ($\phi3$, 0) may lead to a huge change of $\hat{\phi}2$, but a huge change of $\phi2$ affects the target ICR very little. For that reason, it is suggested to decrease the feedback force for steering input $\phi2$ exponentially when the current ICR is approaching the origin of the ICR axis.

As a result, the force feedback errors of the steering interface inputs $\phi 1$ and $\phi 2$ are:

$$e_{\phi 1} = \hat{\phi}1 - \phi 1 \tag{3.26}$$

$$e_{\phi 2} = \left(\hat{\phi}2 - \phi 2\right)\left(1 - e^{-k|\cot(\hat{\phi}1)|}\right) \tag{3.27}$$

where k is a decay constant.

3.3.3 Motor Torque Control

The torque generated by the motor is directly proportional to the current: $T_m = K_m i$. The motor model in (3.8) can be simplified as an RL circuit; hence

$$i = (V - K_b\omega_m)\frac{1}{R}\left(1 - e^{-(R/L)t}\right) \tag{3.28}$$

Assume the time constant R/L is very large; the torque generated by the motor can be simplified to

$$T_m = \frac{K_m}{R}(V - K_b\omega_m) \tag{3.29}$$

Hence we can generate the required torque τ^t by following input voltage:

$$V = \frac{R}{K_m}\tau^t + K_b\omega_m \tag{3.30}$$

3.3.4 Structure of Force Feedback Controller

The feedback force is directly proportional to the force feedback reference $e_{\phi i}$ where $i = 1, 2$. Hence, the required torque becomes

$$\tau^t = K_{\phi i}e_{\phi i} + K_f\omega_m \tag{3.31}$$

where $K_{\phi i}$ is proportional gain. The term $K_f\omega_m$ is added to compensate for the viscous frictional force. Substitute (3.31) into (3.30):

$$V = \frac{RK_{\phi i}}{K_m}e_{\phi i} + \left(\frac{RK_f}{K_m} + K_b\right)\omega_m \tag{3.32}$$

The proportional gain $K_{\phi i}$ can be adjusted according to the requirements. It controls the tolerance of the synchronization. Increasing the gain $K_{\phi i}$ can reduce the synchronization error, but it also reduces the response of the steering motion.

3.4 Behavior-Based Steering Controller

Keeping wheels in minimum lateral slip is a crucial task in steering control as the wheel slip will generate undesired lateral force on wheels that will increase the mechanical stress and induce an amount of driving resistance. It consumes extra driving energy and decreases the vehicle stability. As a result, the proposed behavior-based steering controller consists of two control behaviors: (1) control the orientations of wheels to go to target orientations, and at the same time (2) keep all the wheels in kinematic constraint. The first behavior is accomplished by the position controller while the second behavior is accomplished by the kinematic constraint controller. The kinematic constraint controller acts as virtual linkage among each wheel to minimize wheel slip.

3.4.1 Position Error

The position controller adopts position error as a reference to control steering actuators to turn the wheels to target orientations. The position error is the difference between target orientation δ_i^T and current orientation δ_i^C:

$$e_i^p = \delta_i^T - \delta_i^C \tag{3.33}$$

3.4.2 Kinematic Constraint Error

To keep the orientations of the wheels under the kinematic constraint of steering, the kinematic constraint error for each wheel e_i^k should be quantified.

In pure translation motion, the current ICR is supposed to be located at infinity. In this situation, the kinematic constraint is that all the orientations of the wheels should be equal. The current translation direction is approximated as the average of the orientations of wheels. Hence, the kinematic constraint error of wheel i is defined as

$$e_i^k = \left(\delta_1^C + \delta_2^C + \delta_3^C + \delta_4^C\right)/4 - \delta_i^C \tag{3.34}$$

In turning motion, the kinematic constraint error of each wheel is defined as an additional turning angle required so that the spinning axis of the wheel intersects at the current ICR (see Fig. 3.6).

In (3.20), when $AE - B^2 \neq 0$, the coordinates of current ICR can be found. Similar to (3.16), the appropriate orientation of each wheel (δ_i^a) to keep the kinematic constraint can be determined as follows:

$$\delta_i^a = \operatorname{atan2}\left[D^c\left(x_i - x_{\mathrm{ICR}}^C\right), -D^c\left(y_i - y_{\mathrm{ICR}}^C\right)\right] \tag{3.35}$$

In this case, the kinematic constraint error for each wheel is defined as

$$e_i^k = \delta_i^a - \delta_i^C \tag{3.36}$$

When $AE - B^2 = 0$, the current ICR cannot be found. It implies all spinning axes are in parallel. In this case, (3.20) is not applicable. It also has to satisfy the kinematic constraint for pure translation. Hence, the kinematic constraint error becomes zero in this situation.

3.4.3 Structure of Behavior-Based Steering Controller

The kinematic constraint controller cannot be used alone to control steering motion, as it does not control the wheels to go to the target orientations commanded by the steering interface. It must cooperate with the position controller. Hence, the behavior-based approach is adopted. Figure 3.7 shows the structure of the proposed behavior-based steering controller. It consists of the position controller (P control for short) and the kinematic constraint controller (K control for short). Each control behavior has its own PID gain. Increasing the gain on the K controller can reduce the misalignment of wheels,

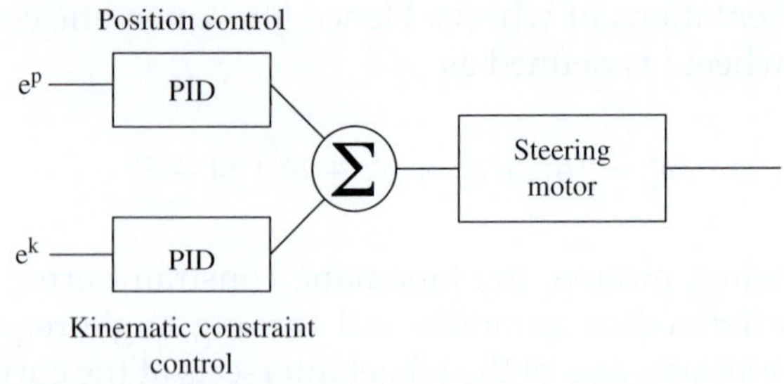

FIGURE 3.7 Block diagram for behavior-based steering controller.

but it also can increase the settling time. As a result, the gain ratio between the P controller and the K controller should be determined according to the requirements of the application. Here we consider that the gain of K control is about 10 times larger than P control, and it can provide satisfactory results.

3.4.4 Stability Analysis

Note from (3.36) that the kinematic constraint controller in turning motion is highly nonlinear. To analyze the stability of the control system, the kinematic constraint controller in turning motion can be linearized as the control in pure translation motion. Equations (3.12) and (3.13) are adopted for stability analysis. The subscript i is added in each variable to distinguish the variables for each wheel. According to the design of the behavior-based steering controller (Fig. 3.7), the control inputs V_i in (3.12) become

$$V_i = K_p^p e_i^p + K_i^p \int e_i^p \, dt + K_d^p \dot{e}_i^p + K_p^k e_i^k + K_i^k \int e_i^k \, dt + K_d^k \dot{e}_i^k \tag{3.37}$$

Assuming there is no disturbance torque, the control system can be represented in state-space form as

$$\dot{X} = AX \tag{3.38}$$

where

$$X = \begin{bmatrix} x_1 & x_2 & \cdots & x_{16} \end{bmatrix}^T \tag{3.39}$$

$$\begin{bmatrix} x_{4(i-1)+1} \\ x_{4(i-1)+2} \\ x_{4(i-1)+3} \\ x_{4(i-1)+4} \end{bmatrix} = \begin{bmatrix} \int \delta_i \, dt \\ \delta_i \\ \delta_i \\ i_i \end{bmatrix} \tag{3.40}$$

$$A = \begin{bmatrix} G & H & H & H \\ H & G & H & H \\ H & H & G & H \\ H & H & H & G \end{bmatrix} \tag{3.41}$$

$$G = \begin{bmatrix} 0 & 1 & 0 & 0 \\ 0 & 0 & 1 & 0 \\ 0 & 0 & \frac{-K_f}{J_m+n^2J_L} & \frac{K_m}{J_m+n^2J_L} \\ \frac{-4K_i^p-3K_i^k}{4} & \frac{-4K_p^p-3K_p^k}{4} & \frac{-4K_d^p-3K_d^k}{4} & \frac{-R}{L} \end{bmatrix} \tag{3.42}$$

$$H = \begin{bmatrix} 0 & 0 & 0 & 0 \\ 0 & 0 & 0 & 0 \\ 0 & 0 & 0 & 0 \\ \frac{K_i^k}{4} & \frac{K_p^k}{4} & \frac{K_d^k}{4} & 0 \end{bmatrix} \tag{3.43}$$

With the use of the parameters listed in Table 3.1, all the eigenvalues of (3.41) have negative real parts. Hence it can be concluded that the control system with the selected PID parameters is stable.

3.5 Simulations and Experiments

For safety reasons, the proposed design and algorithms of the extended steering interface must be verified in a virtual environment before implementation on the real vehicle. It is accomplished by HIL simulations. In the following simulations, the Real-Time Windows Target[38] in MATLAB is selected as the HIL simulation environment. To realize the steering effect, the Virtual Reality Toolbox[37] is used to visualize the vehicle motion.

3.5.1 Structure of Simulation Environment

Figure 3.9 shows the whole HIL simulation platform while the architecture of HIL simulation can be found in Fig. 3.8. It consists of a host PC and real hardware. Communication between the host PC and real hardware is done over a RS232 serial line. The host PC acts as a vehicle control system that communicates with the extended steering interface and receives the signal from the foot pedal. It calculates the target orientations of the wheels and appropriate steering interface posture according to the information acquired from the extended steering interface and the dynamics model, respectively. It also acts as a steering motor controller. The host PC simulates the response of steering motors and vehicle body to evaluate the proposed force feedback controller and steering interface. The motion of the vehicle and wheels can be visualized in the virtual reality environment. Real hardware in the simulation includes the extended steering interface and a foot pedal. In the extended steering interface,

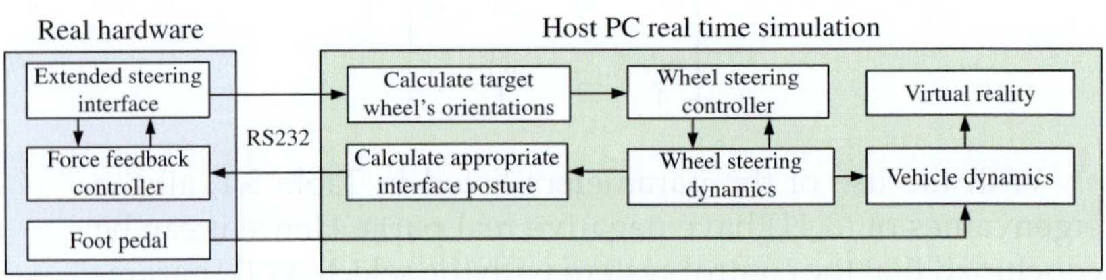

FIGURE 3.8 Structure of hardware-in-the-loop (HIL) simulation.

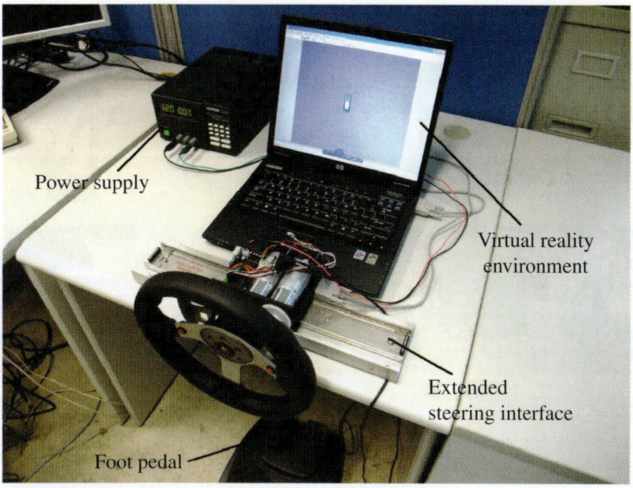

FIGURE 3.9 HIL simulation platform.

the steering input $\phi 3$ is realized virtually in the host PC. The foot pedal is introduced to define the longitudinal driving force of each wheel. Table 3.1 clarifies the notations and values used in the modeling and simulations.

3.5.2 Steering Motion Simulation

In the HIL simulation, the authors tried to steer the vehicle with different steering inputs. Figure 3.10 shows some of the simulation results with $\phi 3 = -1.25$. All the steering motions are correct. The ICR can be changed to any point continuously. The extended steering interface provides driver with omnidirectional controllability. Driver can experience a novel and powerful driving style.

3.5.3 Force Feedback Simulation

In the force feedback simulation, driver tried to change the posture of the steering interface rapidly and randomly. Figure 3.11 shows the simulation results with and without activating the force feedback controller. The actual and appropriate steering postures are marked as blue and red, respectively. It can be

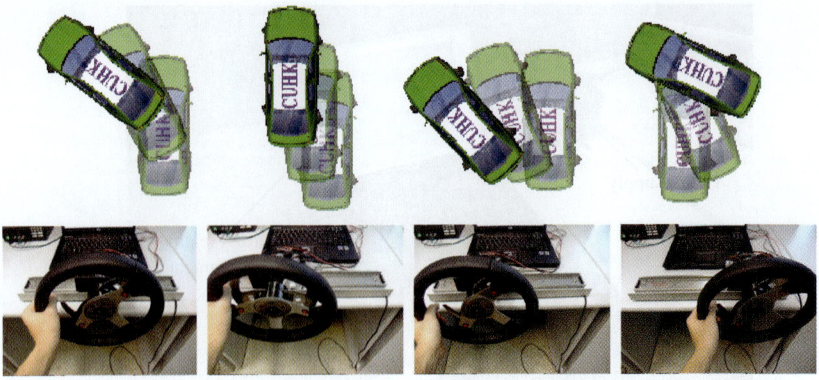

FIGURE 3.10 Examples of steering results.

observed that both results have certain time delay (around 0.2 s) between actual and appropriate interface posture. The curvatures between actual and appropriate interface posture by using the force feedback control are more alike than not using the force feedback control. When the time delay is compensated,

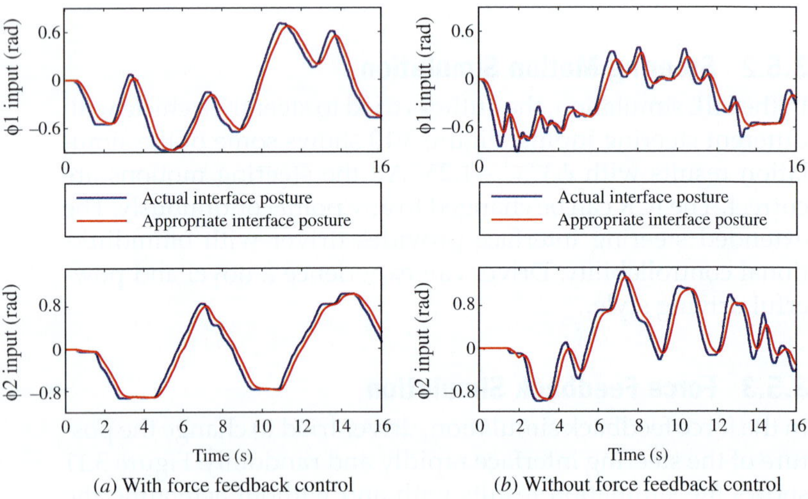

FIGURE 3.11 Simulation results of force feedback control.

the maximum difference between actual and appropriate interface posture with and without using the force feedback control are 0.09*rad* and 0.2*rad*, respectively. It indicates that the force feedback controller succeeds to reflect the wheel conditions to driver through the extended steering interface.

3.5.4 Steering Behavior

This simulation is implemented to realize how the behavior-based controller acts to reduce wheel slip under disturbance. It is assumed that all the driving actuators provide uniform constant torques, that is, 250 N·m. The initial speed of the vehicle and the orientations of the wheels are zero. Vehicle is commanded to drive in turning motion about fixed ICR (0, 10). A pulse torque is applied on wheel 1 at 2 s to imitate an environmental disturbance. The simulation results are shown in Fig. 3.12. When the position controller is used, the disturbance changes only the orientation of wheel 1. However, when the

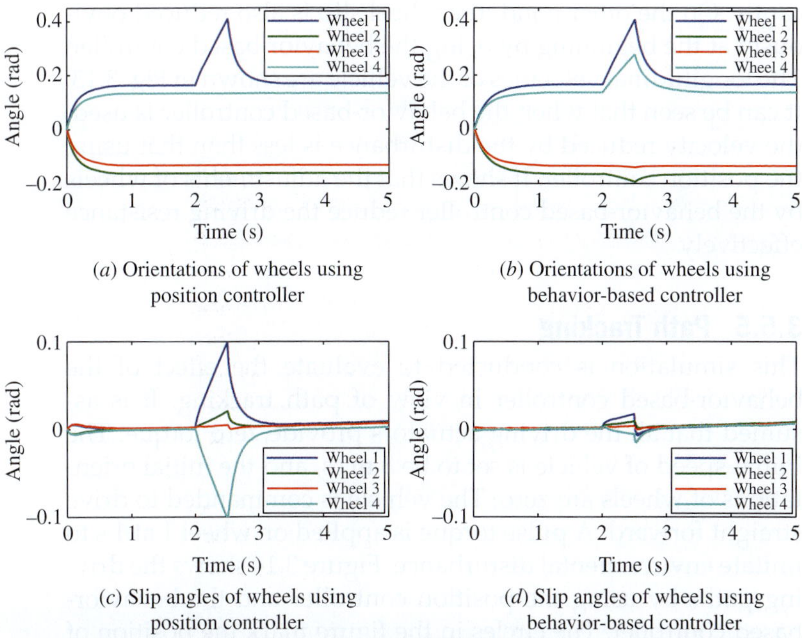

(*a*) Orientations of wheels using position controller

(*b*) Orientations of wheels using behavior-based controller

(*c*) Slip angles of wheels using position controller

(*d*) Slip angles of wheels using behavior-based controller

FIGURE 3.12 Simulation results of steering control.

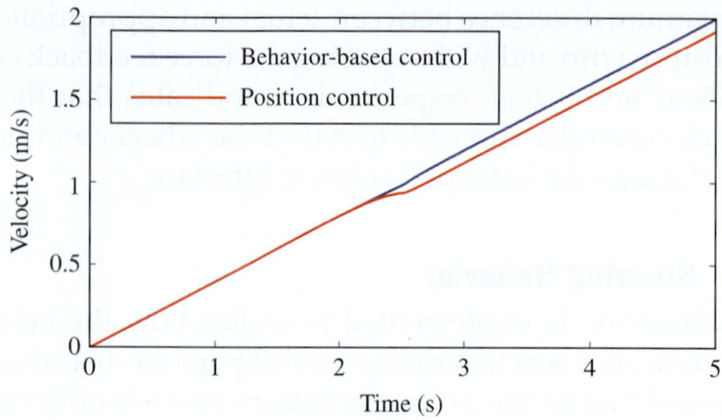

FIGURE 3.13 Longitudinal velocities by using position controller and behavior-based controller.

behavior-based controller is used the orientations of the other wheels have corresponding adjustments (Fig. 3.12*b*). It can be realized by comparison of Fig. 3.12*c* and *d* that the corresponding adjustments result in reduction of the wheel slip significantly. On the other hand, the wheel slip is also reduced obviously at the beginning by using the behavior-based controller. The longitudinal velocities of the vehicle are shown in Fig. 3.13. It can be seen that when the behavior-based controller is used, the velocity reduced by the disturbance is less than that using the position controller. It shows that the adjustments of wheels by the behavior-based controller reduce the driving resistance effectively.

3.5.5 Path Tracking

This simulation is conducted to evaluate the effect of the behavior-based controller in view of path tracking. It is assumed that all the driving actuators provide zero torque. The initial speed of vehicle is set to be 3 m/s, and the initial orientations of wheels are zero. The vehicle is commanded to drive straight forward. A pulse torque is applied on wheel 1 at 1 s to imitate environmental disturbance. Figure 3.14 shows the driving paths by using the position controller and the behavior-based controller. The circles in the figure mark the position of vehicles in every 2 s. When disturbance is applied, the path varies a lot by using the behavior-based controller, whereas

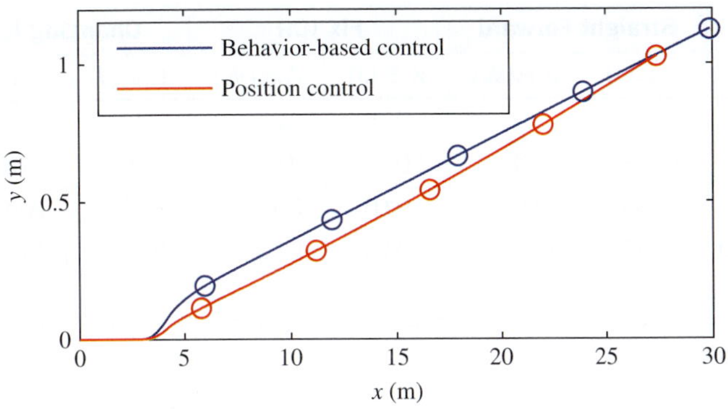

FIGURE 3.14 Driving paths by using position controller and behavior-based controller.

the variation is soon equal to that using the position controller. In addition, it shows that the vehicle can run much farther by using the behavior-based controller. It shows that using the behavior-based controller can improve the driving efficiency, but the cost is the small variation of driving path. In fact, in a nonautonomous vehicle, it is not a serious problem as it does not need to follow the predefined path precisely. Furthermore, the variation of path can easily be corrected by the driver.

3.5.6 Driving Efficiency

To evaluate the performance of the proposed behavior-based controller in view of driving efficiency, two indexes are recorded in the following simulations: the kinetic energy (K.E.) of vehicle at the end of simulations and the average power (A.P.) applied on the steering motors. The simulations are implemented in three different steering commands: straight forward, fix ICR (0,10), and periodic changing of ICR [swap between (0,−3) and (0,3)]. All the initial orientations of the wheels are zero. It is assumed that all the driving actuators provide uniform constant torques, that is, 250 N·m. Random disturbances on the wheels are applied in several simulations. Table 3.2 shows the simulation results. The notations P and B represent the use of the position controller and the behavior-based controller, respectively; +D represents the random disturbances applied on each wheel. The results show that in all kinds of steering commands, when using the

	Straight Forward		Fix ICR		Changing ICR	
	K.E.(J)	A.P.(W)	K.E.(J)	A.P.(W)	K.E.(J)	A.P.(W)
P	1008	0	925	525	937	761
B	1008	0	1022	485	978	661
P+D	993	115	916	544	886	838
B+D	1000	164	1013	513	970	744

TABLE 3.2 Table of Kinetic Energy and A.P. Values in Different Steering Commands

behavior-based controller, the total kinetic energy is larger. It implies the behavior-based controller is efficient to reduce the driving resistance in many cases. As for the average steering power, except for the case of driving straight forward, the values by using the behavior-based controller are smaller than those using the position controller. It is inferred that using the behavior-based controller may even reduce the energy consumption for steering. The results indicate that by using the behavior-based controller, the higher kinematic energy of the vehicle is not caused by increasing the steering power. In summary, it can improve both driving and steering efficiency.

3.6 Summary

This chapter presents an omnidirectional steer-by-wire system for a nonautonomous 4WIS vehicle. The system includes the extended steering interface to the behavior-based steering controller. By using the extended steering interface, drivers can steer the 4WIS vehicle in the traditional way or about an ICR without any mode-switching operation. The steering interface also includes the force feedback function to reflect the wheel conditions to the driver. The hardware prototype of the steering interface is developed to verify the proposed design. HIL simulations are carried out to validate the correctness of the proposed steering interface and the force feedback control algorithm. The behavior-based steering controller consists of the position controller and kinematic constraint controller. The kinematic constraint controller acts as a virtual linkage among the wheels to maintain kinematic constraint so as to minimize wheel slip. Using this approach, no trajectory planning is needed. It is beneficial for nonautonomous vehicles that the driving path cannot be predefined by the control system. A number of simulations are conducted to examine the performance of the proposed controller. The simulation results show that when compared with a position controller, the behavior-based controller improves both the driving and the steering efficiency with a little path variation.

CHAPTER 4

Four-Wheel Independent Force Control

This chapter proposes the four-wheel independent force control. Section 4.1 introduces traction force distribution, including the deterministic force generation and optimal traction force distribution. Section 4.2 presents the proposed direct yaw moment control (DYC) including the wheel-slip observer, yaw rate error observer, and heading-error observer. Numerical simulations results are presented for both sections.

4.1 Traction Force Distribution

The 4WID and 4WIS electric vehicles are an innovative configuration.[75] In the 4WIS configuration, the orientations of the wheels can be controlled independently. It enables the vehicle to be steered omnidirectionally. In the 4WID configuration, each wheel is driven by an in-wheel motor independently. As a result, the traction force on each wheel can be controlled independently. The independent control of traction force is beneficial for optimizing the driving efficiency[115] and improving vehicle stability.[75, 95, 104]

The vehicle *stability* is defined as the ability to achieve a given curvature (turning radius). DYC is the most common technique for vehicle stability control in the 4WID vehicle.[22, 34] It relies on the feedback information of the vehicle motion. However, the accurate vehicle motion is difficult to obtain. It makes the DYC difficult to implement. Without using the information of vehicle motion, the most common and simplest traction force distribution on the 4WID vehicle is to distribute uniformly, whereas the uniform force distribution is not the best solution in view of vehicle stability. Studies found that the yaw moment generated by uniform force distribution is always more than required when cornering (see Sec. 4.1.3). The extra yaw moment always forces the vehicle out of a given curvature and hence results in oversteer.

As a result, this chapter proposes an optimal traction force distribution for the 4WID-4WIS vehicle to improve vehicle stability without using any feedback information of vehicle motion. It is accomplished by using the proposed deterministic force generation algorithm. The deterministic force generation algorithm determines the traction force on each wheel to generate the yaw moment and lateral force on the vehicle body deterministically. The algorithm depends on only the location of the center of mass of the vehicle body, locations of wheels, and driver control inputs such as steering angle and acceleration pedal. Those are easy to obtain, and no parameter tuning is needed. Hence the proposed algorithm can be implemented easily. Although the proposed force distribution and deterministic force generation algorithms are designed for the 4WID-4WIS vehicle, they can also be applied to any steering types of 4WID vehicle such as typical 2WS, as 4WIS is the general case of any steering configuration.

Figure 4.1 shows the notations of a 4WID-4WIS vehicle model used in this chapter. Since only the planer motion is considered, the roll, pitch, and vertical motion are neglected. The origin of the coordinate system is located at the center of

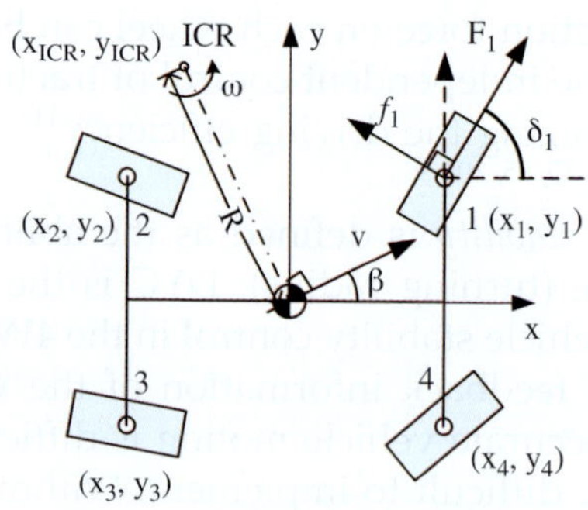

FIGURE 4.1 Notations of the 4WID-4WIS vehicle model.

mass of the vehicle body. ICR denotes the instantaneous center of rotation. δ_i is the orientation of wheel i, the values can be determined by a target ICR. R is the turning radius. v and ω represent the tangential and angular velocity, respectively. β denotes the traveling angle of the center of mass. F_i and f_i denote the longitudinal (traction) and lateral force on wheel, respectively. The tangential direction is defined as the traveling direction of the center of mass while the centripetal direction is defined as the direction toward the ICR.

4.1.1 Deterministic Force Generation

The yaw moment and centripetal force provided by longitudinal traction force can be controlled deterministically. To generate the required yaw moment and centripetal force ε_m and ε_n, the traction force should be arranged such that

$$\text{Centripetal force: } \sum_{i=1}^{4} b_i F_i = \varepsilon_n \tag{4.1}$$

$$\text{Yaw moment: } \sum_{i=1}^{4} c_i F_i = \varepsilon_m \tag{4.2}$$

where $b_i = \sin(\delta_i - \beta)$ and $c_i = x_i \sin \delta_i - y_i \cos \delta_i$. The resultant tangential force ε_t can be written as

$$\text{Tangential force: } \sum_{i=1}^{4} a_i F_i = \varepsilon_t \tag{4.3}$$

where $a_i = \cos(\delta_i - \beta)$. Furthermore, the total traction force is determined by the acceleration pedal:

$$\sum_{i=1}^{4} |F_i| = F_{\text{total}} \tag{4.4}$$

where F_{total} is the total traction force proportional to the angle of the acceleration pedal. To simplify constraint (4.4) and provide proper tangential acceleration, the traction forces are further constrained as

$$F_i \geq 0 \qquad \forall i = 1, 2, 3, 4 \tag{4.5}$$

By combining (4.1) to (4.5), the force distribution can be formulated as

$$\begin{bmatrix} F_1 \\ F_2 \\ F_3 \\ F_4 \end{bmatrix} = A^{-1} \begin{bmatrix} \varepsilon_t \\ \varepsilon_n \\ \varepsilon_m \\ F_{\text{total}} \end{bmatrix} \tag{4.6}$$

where $A = \begin{bmatrix} a_1 & a_2 & a_3 & a_4 \\ b_1 & b_2 & b_3 & b_4 \\ c_1 & c_2 & c_3 & c_4 \\ 1 & 1 & 1 & 1 \end{bmatrix}$.

Since ε_t is not fixed by requirement, there are multiple solutions of the force distribution. However, ε_t cannot be freely assigned. To fulfill constraint (4.5), ε_t must be chosen such that $F_i \geq 0$ for all i. Figure 4.2 shows the admissible range of ε_t with following parameters: $F_{\text{total}} = 1000$ N, ICR = $(-1.25, 6.31)$,

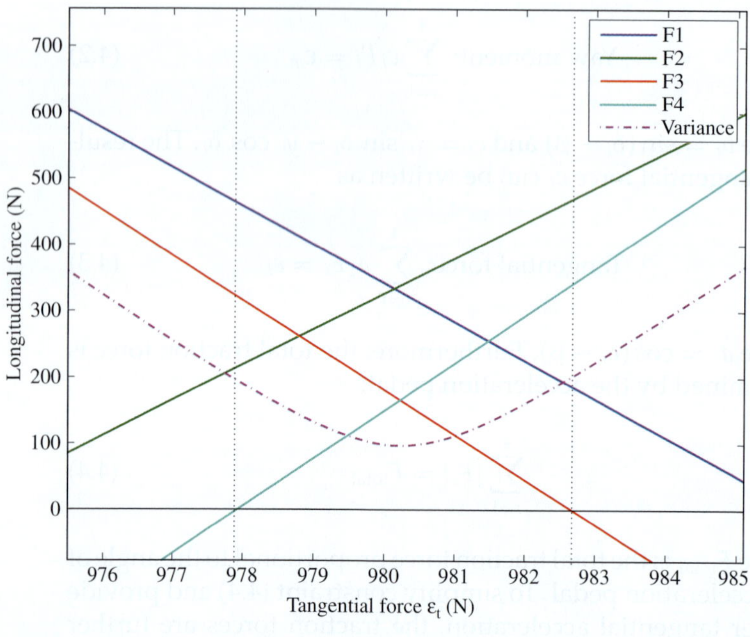

FIGURE 4.2 Admissible force distribution.

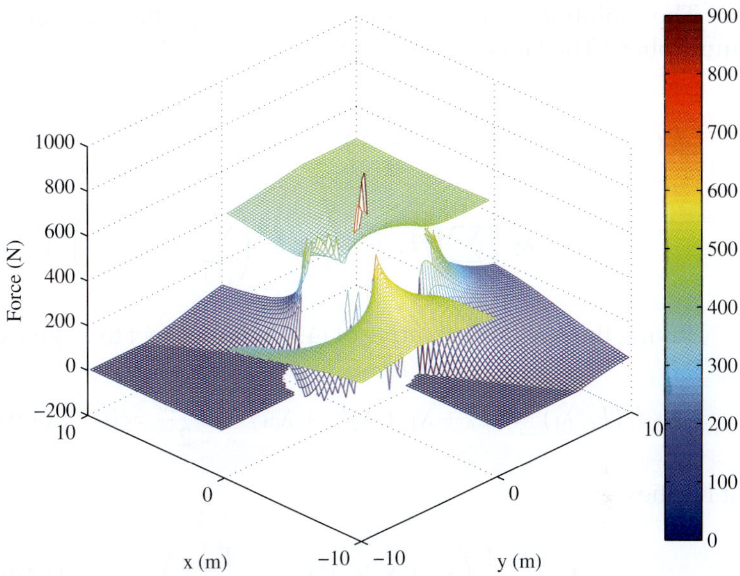

FIGURE 4.3 Force distribution of method I.

$\varepsilon_m = 0$; and $\varepsilon_n = 0$. In this case, ε_t is restricted between 977.9 N and 982.7 N. To maximize the driving efficiency, ε_t should be chosen as large as possible. This selection criterion is denoted as method I. Figure 4.3 shows the force distributions on wheel 1 with different ICR by method I. In the figure, x and y coordinates indicate the location of ICR. It can be seen that the force is discontinuous among different quadrants, and the variance is very large. The sudden large change of torque may damage the in-wheel motor. In addition, the large traction force may result in serious wheel slip and hence may reduce vehicle stability. The alternative criterion, which is denoted as method II, is to choose ε_t such that the force variance among each wheel is minimum:

$$\text{Minimize} \sum_{i=1}^{4} \left(F_i - \frac{F_{\text{total}}}{4} \right)^2 \tag{4.7}$$

The solution can be found by using the Lagrange multiplier.[5] The lagrangian function is then written as

$$L(F_i, \lambda_i) = \sum_{i=1}^{4}\left(F_i - \frac{F_{\text{total}}}{4}\right)^2 + \lambda_1\left(\sum_{i=1}^{4} F_i - F_{\text{total}}\right)$$
$$+\lambda_2\left(\sum_{i=1}^{4} b_i F_i - \varepsilon_n\right) + \lambda_3\left(\sum_{i=1}^{4} c_i F_i - \varepsilon_m\right) \quad (4.8)$$

Taking the partial derivative of (4.8) with respect to F_i gives

$$\Delta_{F_i} L(F_i, \lambda_i) = 2F_i + \lambda_1 + \lambda_2 b_i + \lambda_3 c_i - \frac{F_{\text{total}}}{2} = 0 \quad (4.9)$$

It implies

$$F_i = -\frac{1}{2}\left(\lambda_1 + \lambda_2 b_i + \lambda_3 c_i - \frac{F_{\text{total}}}{2}\right) \quad (4.10)$$

Substitute (4.10) into (4.1), (4.2), and (4.4):

$$\begin{bmatrix} 4 & \sum b_i & \sum c_i \\ \sum b_i & \sum b_i^2 & \sum b_i c_i \\ \sum c_i & \sum b_i c_i & \sum c_i^2 \end{bmatrix} \begin{bmatrix} \lambda_1 \\ \lambda_2 \\ \lambda_3 \end{bmatrix} = \begin{bmatrix} 0 \\ \frac{F_{\text{total}}}{2}\sum b_i - 2\varepsilon_n \\ \frac{F_{\text{total}}}{2}\sum c_i - 2\varepsilon_m \end{bmatrix} \quad (4.11)$$

And λ_i can be written in terms of ε_m and ε_n:

$$\lambda_i = o_i + n_i \varepsilon_n + m_i \varepsilon_m \quad (4.12)$$

where m_i, n_i, and o_i can be deduced from (4.11) using the inverse matrix method. Substitute (4.12) into (4.10), and F_i can be formulated in terms of ε_m and ε_n:

$$F_i = O_i + N_i \varepsilon_n + M_i \varepsilon_m \quad (4.13)$$

where M_i, N_i, and O_i can be deduced from (4.10) and (4.12).

Figure 4.4 shows the force distributions on wheel 1 with different ICR by method II. The force is continuous among different ICR, and the variation is relatively small. However, the resultant tangential force reduced too. Figure 4.5 shows the

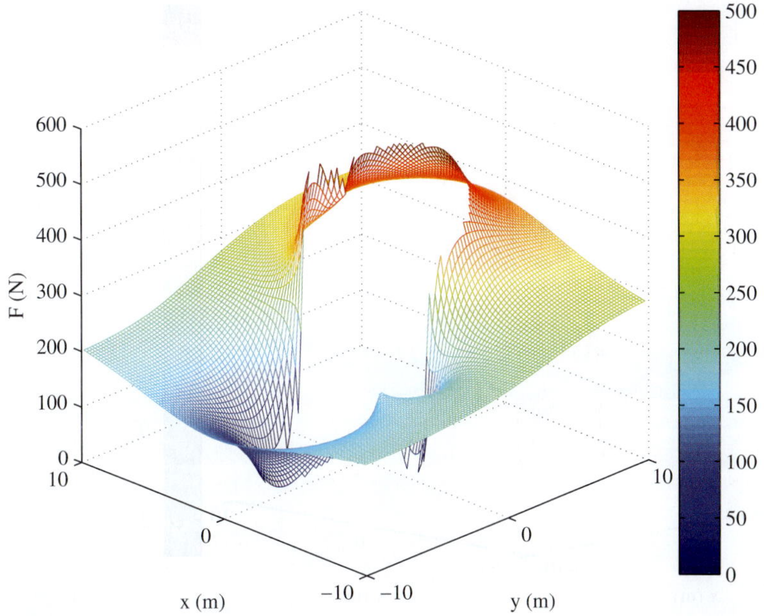

FIGURE 4.4 Force distribution of method II.

percentage difference of ε_t between method I and method II. The difference is less then 1 percent when the turning radius is larger than 1 m. The difference will further decrease when the turning radius increases. It is worth it to sacrifice less than 1 percent of tangential force to further improve the vehicle stability.

Although the proposed force generation algorithm can generate the required yaw moment and centripetal force ε_m and ε_n deterministically, the values of ε_m and ε_n are confined to a certain range according to the orientation of the wheels and the total traction force. To fulfill constraint (4.5), ε_m and ε_n should be selected such that

$$O_i + N_i\varepsilon_n + M_i\varepsilon_m \geq 0 \qquad \forall i = 1, 2, 3, 4 \tag{4.14}$$

With these four inequality constraints, the inequality constraint graph can be constructed with respect to ε_m and ε_n. Figure 4.6 shows the inequality constraint graph with the

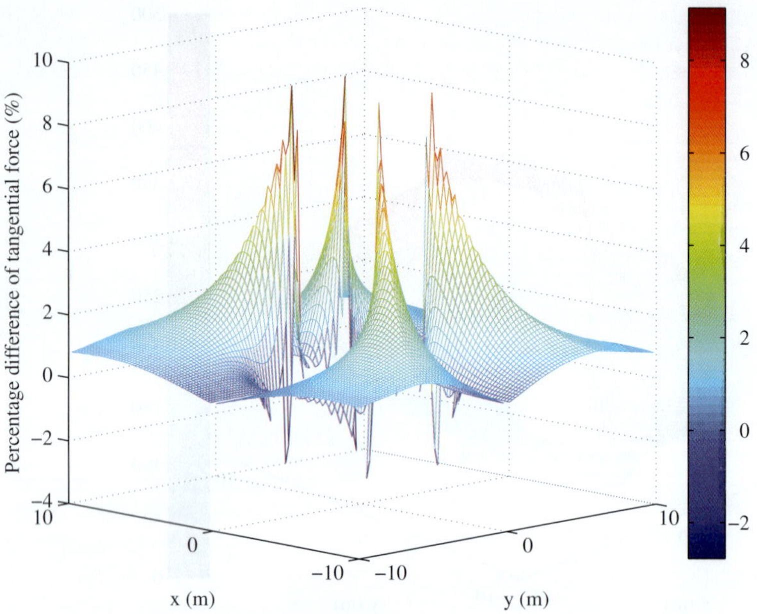

FIGURE 4.5 Difference of ε_t between method I and method II.

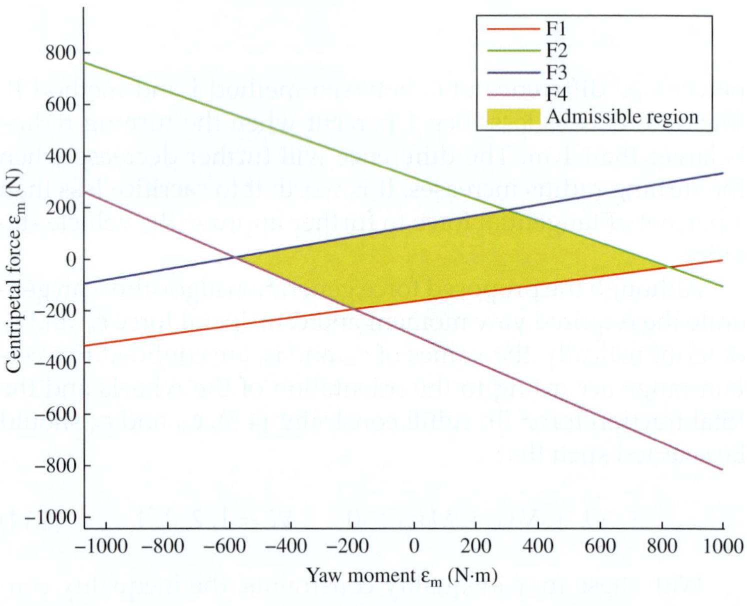

FIGURE 4.6 Inequality constraint graph and admissible region.

following parameters: $F_{total} = 1000$ N, ICR = (−1.25, 6.31). The region colored in yellow represents the admissible yaw moment and lateral force.

4.1.2 Optimal Traction Force Distribution

The vehicle *stability* is defined as the ability to achieve a given curvature. To follow the given curvature, the resultant yaw moment and centripetal force must equal certain values according to the vehicle velocity and angular acceleration. In the 4WID vehicle, the most common force distribution is to distribute uniformly. When cornering, the uniform force distribution does not generate only the tangential force, but also the yaw moment. Figure 4.7 shows the yaw moments generated by uniform force distribution with different ICR. When the angular acceleration of a vehicle is small, the generated yaw moment is always more than required and hence results in oversteer. If the vehicle speed exceeds the critical speed, the terminal oversteer occurred. The vehicle loses control and turns much more than requested. As

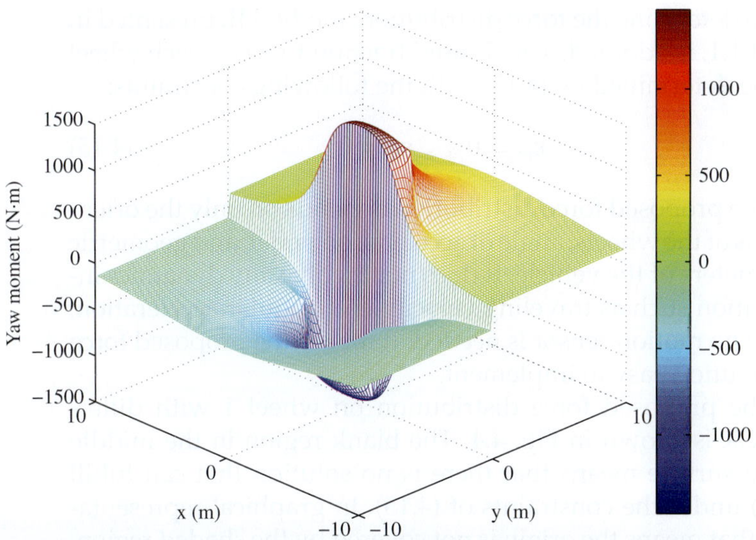

FIGURE 4.7 Yaw moment generated by uniform force distribution.

a result, the uniform force distribution is not a good option in view of vehicle stability.

It is realized in Sec. 3.1 that the vehicle dynamics are highly nonlinear. Although the longitudinal tire force can be controlled directly, it is not the case for the lateral tire force. It is also difficult to measure the lateral tire force. To keep the vehicle in a given turning curvature, one approach is to avoid any lateral slip of the tire. It can be achieved by providing all the required yaw moment and centripetal force by longitudinal tire forces. However, as mentioned in Sec. 4.1.1, the generation of yaw moment and centripetal force by longitudinal tire force is limited. This approach is feasible only in a certain range of vehicle speed and angular acceleration. It is impossible to avoid any lateral slip of wheel in high-speed cornering. As a result, the proposed force distribution is aimed at providing a stable condition to minimize the path following error.

It is known that both neutral steer and understeer are said to be stable[43] whereas neutral steer has a higher tendency to follow the given curvature. Hence it is aimed to distribute the force to reach neutral steer. It can be achieved by distributing the traction force to provide zero yaw moment and centripetal force.

To determine the force distribution, method II, presented in Sec. 4.1.1, is adopted. The desired traction force on each wheel can be determined by (4.13) with the following constraints:

$$\varepsilon_m = 0 \qquad \varepsilon_n = 0 \tag{4.15}$$

The proposed force distribution depends on only the orientations of the wheels, angle of acceleration pedal, and geometric parameters of the vehicle. It does not involve any dynamics information such as traveling velocity and angular acceleration; hence no motion sensor is needed. It makes the proposed force distribution easy to implement.

The proposed force distribution on wheel 1 with different ICR is shown in Fig. 4.4. The blank region in the middle of the surface means that there is no solution that can fulfill (4.14) under the constraints of (4.15). In graphical representation, that means the origin is not covered by the shaded region. In case there is no solution that fulfills the constraint (4.14), ε_m

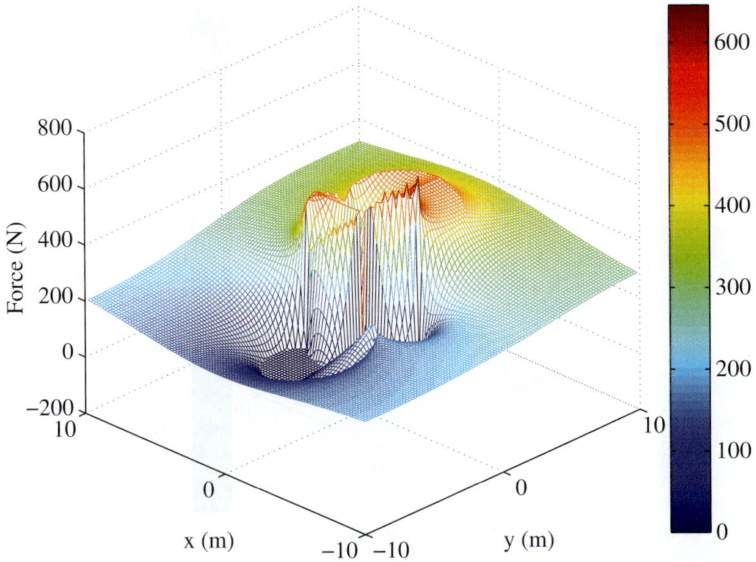

FIGURE 4.8 Complete force distribution.

is modified as a value in the shaded region that is closest to the origin. The proper ε_m can be found by evaluating the inequality constraint graph constructed by (4.14). Figure 4.8 shows the complete force distributions of wheel 1, and Fig. 4.9 shows the generated yaw moments. The yaw moments are always less than those generated by uniform force distribution.

4.1.3 Performance Analysis

The simulations are implemented to evaluate the performance of the proposed force distribution. The results of using the uniform force distribution are also presented for comparison. The evaluations are divided in two parts: vehicle stability and driving efficiency. The mathematical models presented in Sec. 3.1 are adopted for simulations. In the simulations, the initial speed and total traction force are set as 0 m/s and 1000 N, respectively. In addition, it is assumed that the traction force F_i can be controlled as required, definitely without any error and delay.

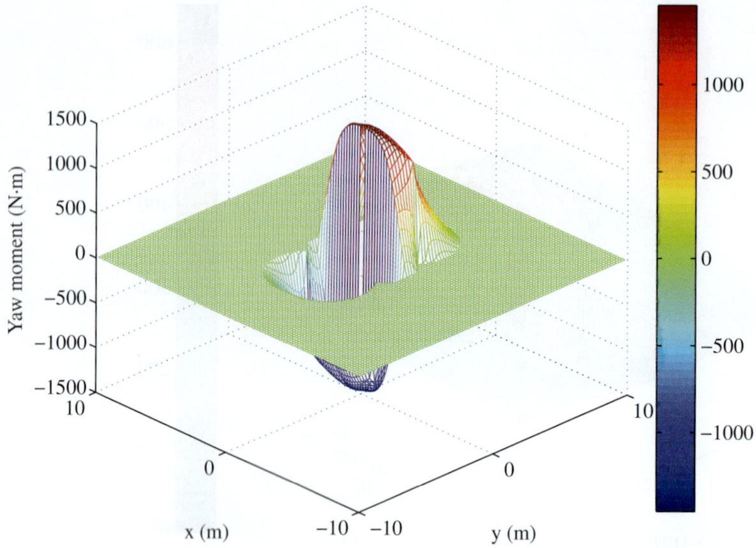

FIGURE 4.9 Yaw moment generated by proposed force distribution.

4.1.3.1 Vehicle Stability

Table 4.1 shows the percentage errors of the turning radius by using the proposed and uniform force distributions in different target turning radii. The percentage errors at the beginning and at the stable state are recorded. The positive error means that the resultant turning radius is larger than the target turning radius and the negative error represents that the resultant turning radius is smaller than the target turning radius. Figure 4.10 shows the turning radius change with time with target turning radius equal to 79.58 m.

Note that at the beginning, the errors from using the proposed force distribution are about 0.44 percent but while using the uniform force distribution are almost zero. The acceleration is high at the beginning. The vehicle requires high yaw moment to support the vehicle angular acceleration. However, the proposed force distribution does not provide yaw moment, and hence the required yaw moment is provided by the tire lateral force. It results in a small amount of understeer.

However, the errors caused by using the proposed force distribution are gradually decreased while the errors caused by

	R (m)	70.03	95.5	127.3	159.2	254.6
Proposed	begin stable	0.44% 0.17%	0.44% 0.02%	0.47% 0.00%	0.44% −0.06%	0.47% 0.00%
Uniform	begin stable	0.01% −2.68%	0.01% −1.42%	0.00% −0.86%	0.00% −0.75%	0.04% −0.51%

TABLE 4.1 Tracking Error in Different Turning Radii

using the uniform force distribution are gradually increased. It is because the vehicle speed is increasing but the acceleration is decreasing due to the air resistance. The required yaw moment to maintain the given curvature is also reducing. Hence the extra yaw moment generated by the uniform force distribution results in oversteer.

In Table 4.1 and Fig. 4.10, note that the errors increase when the target turning radius decreases by using both force distributions. However, the errors from using the uniform force

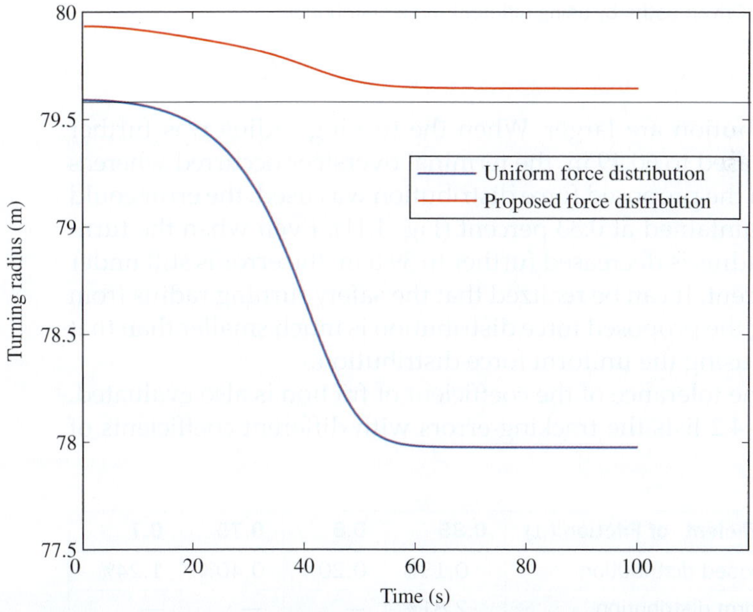

FIGURE 4.10 The change in turning radius from using different force distributions.

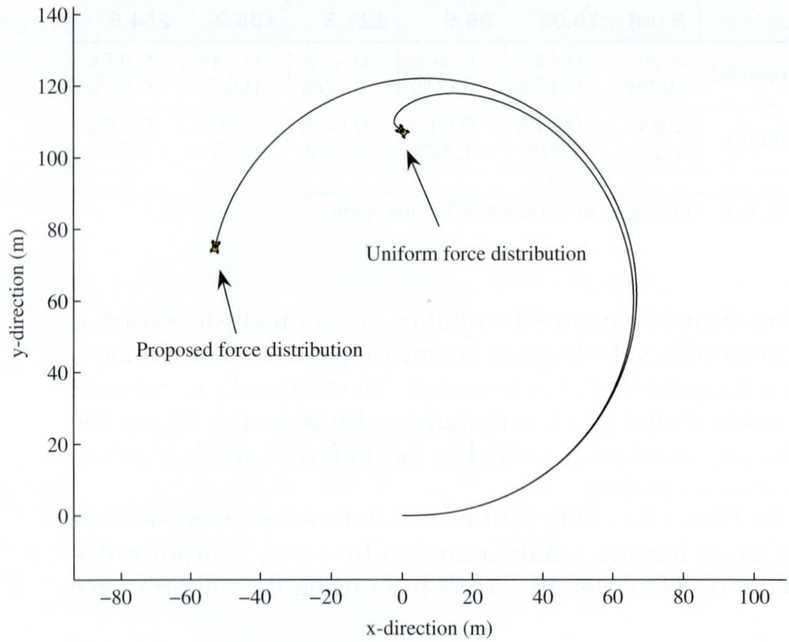

FIGURE 4.11 Driven paths by using different force distributions.

distribution are larger. When the turning radius was further decreased to 60.49 m, the terminal oversteer occurred whereas when the proposed force distribution was used, the error could be maintained at 0.33 percent (Fig. 4.11). Even when the turning radius is decreased further to 39.8 m, the error is still under 3 percent. It can be realized that the safety turning radius from using the proposed force distribution is much smaller than that from using the uniform force distribution.

The tolerance of the coefficient of friction is also evaluated. Table 4.2 lists the tracking errors with different coefficients of

Coefficient of Friction (μ)	0.85	0.8	0.75	0.7
Proposed distribution	0.13%	0.20%	0.40%	1.24%
Uniform distribution	−2.81%	—	—	—

TABLE 4.2 Tracking Error in Different Coefficients of Friction

Actual Turning Radius (m)	77.98	94.14	126.2	189.8
Proposed v (m/s)	25.699	27.554	30.076	32.584
Uniform v (m/s)	25.700	27.555	30.075	32.585

TABLE 4.3 Vehicle Top Speed in Different Turning Radii

friction. The target turning radius is fixed as 79.58 m. By using the uniform force distribution, the terminal oversteer occurs when the coefficient of friction is 0.8. Nevertheless, when the proposed force distribution is used, the vehicle can keep track of the given curvature with small error (0.20 percent). Even when the coefficient of friction is further decreased to 0.7, the error is just 1.24 percent. It indicates that the proposed force distribution can also improve the vehicle stability in slippery road conditions.

4.1.3.2 Driving Efficiency

To evaluate the driving efficiency, the terminal speeds with different turning radii are recorded as shown in Table 4.3. It is discovered that the terminal speeds by both force distributions are almost equal. The differences are extremely small. It can be concluded that the proposed force distribution does not decrease the driving efficiency.

4.2 Direct Yaw Moment Control

DYC is one of the most common techniques for enhancing vehicle controllability and stability. In the latest technology on traditional ICE driven vehicles, the required yaw moment to stabilize a vehicle is generated by an individual control of a wheel braking force through brake discs, which is known as the differential braking method.[7, 79, 83] In recent decades, electric vehicles have become popular. It is reported that the better response of electric vehicles powered by an electric motor can generate traction or braking forces more precisely. An electric in-wheel motor installed in each wheel is one of the common configurations on electric vehicles, which is known as 4WID. It enables the longitudinal traction force on each wheel to be controlled independently. The independent control of the traction force is beneficial for optimizing the driving efficiency[115] and improving vehicle stability.[74, 95, 104] Based on these structural merits, 4WID vehicles can generate the yaw moment precisely by the differential driving method. This method is widely used in the literature such as Refs. 22, 25, 34, 46, 65, 94, and 99.

Vehicle stability is defined as an ability to follow a target trajectory. It mainly includes two references for tracking, i.e., yaw rate and heading angle. Canale et al.[11] proposed a DYC by compensating only the yaw rate error. The yaw rate error can be obtained easily by comparing the steering input and the vehicle yaw rate measured by gyroscope. However, tracking only the yaw rate may induce accumulative heading error and make the vehicle out of control when the accumulated heading error is sufficiently large. As a result, the authors of Refs. 27, 64, and 109 proposed their DYC by tracking both yaw rate and heading angle. In their methods, the heading angle is defined as the side slip angle of a wheel. This assumption is correct when the steering angle is small. Because the real-time measurement of the side slip angle is expensive, which is impractical in application, many researchers, such as Kim and Kim,[46] proposed using a tire model to estimate the side slip angle. The estimation is accurate when the parameters of the tire model and road surface such as stiffness and frictional coefficient are known and the tire dynamics is assumed to be in a linear region. In practice, the conditions of the wheels and road surface can be time-varying. It thus restricts the application to a constant driving condition.

In addition, most of the works are just focused on traditional front-wheel steering vehicles, and a small steering angle is assumed. The stability issue of 4WIS vehicles that have omnidirectional steerability has not been discussed. As the 4WIS configuration allows high maneuverability, it is one of the popular configurations in wheeled mobile robots (see, e.g., Lauria et al.[50]). In recent decades, a number of 4WIS vehicles were developed such as the Toyota Fine-T, Nissan Pivo2, and OK-1. It is believed that the 4WIS configuration will become a popular setting in future electric vehicles.

In view of the limitations of the current DYC technology, we propose a novel DYC for 4WID-4WIS vehicles. The desired yaw moment is generated by a deterministic force distribution method (Sec. 4.1) that generates the yaw moment and centripetal force deterministically by longitudinal tire force in any steering command. In the proposed yaw moment controller, it tracks both the yaw rate and the heading angle to enhance vehicle steerability and stability, respectively. Different from the conventional side slip approximation method, the proposed heading-error observer is independent of the characteristics of the tires and road surface. It makes the observer applicable to any road condition. Although the works are focused on 4WID-4WIS vehicles (see Fig. 4.2), the proposed heading-error observer can be generally applied to other types of vehicles. The proposed DYC and wheel-slip observer can also be adapted to different types of 4WID vehicles.

4.2.1 Deterministic Yaw Moment Control

The main functions of the proposed yaw moment controller are to enhance vehicle steerability by tracking the desired yaw rate and at the same time to suppress the heading error in the stable region to stabilize the vehicle. The proposed controller includes feedforward and feedback control. The feedforward controller generates a yaw moment to turn the vehicle according to the steering inputs and longitudinal acceleration of the vehicle body. Meanwhile the feedback controller generates the desired yaw moment to minimize the yaw rate and heading errors. As a result, the yaw rate error observer and heading-error observer are proposed and adopted in the system. In addition, the system requires the tangential velocity of the vehicle to be known, and it is estimated by a spinning speed of a wheel. For that reason, the proposed controller is activated when the amount of longitudinal wheel slip is small. The wheel slip is detected by the proposed wheel-slip observer. When serious wheel slip occurs, an antislip controller dominates the control to eliminate the slipping effect. There are many antislip controllers proposed in the literature.[25] As this chapter is mainly focused on the yaw moment controller, the discussion of the antislip control is skipped.

4.2.1.1 Wheel-Slip Observer

In conventional wheel-slip observers, only one wheel is considered and a zero steering angle is assumed. It is not enough to describe the slipping situation in 4WID-4WIS vehicles. As a result, a wheel-slip observer for 4WID-4WIS vehicles is proposed that considers all wheels and omnidirectional steering. A wheel slip in a 4WID-4WIS vehicle can be detected by analyzing the dynamics of the wheels and the vehicle body:

$$\sum \left[\left(\tau_i - J\,\frac{d\omega_i}{dt} \right) \frac{R_i}{r} \right] + RF_R = \left(I + mR^2 \right) \frac{d\gamma}{dt} \tag{4.16}$$

where τ_i is the torque generated by the in-wheel motor in wheel i; r is the radius of a wheel that is assumed to be identical and constant in all wheels; ω_i is the spinning speed of wheel i; J and I are the moment of inertia of a wheel and the vehicle body about the center of mass, respectively; m is the mass of the vehicle; γ is the yaw rate of the vehicle body; R_i is the distance between the center of wheel i and the target ICR; and F_R represents the frictional forces induced by the internal structure and environment such as aero force.

When slipping occurs on wheel j,

$$\frac{r}{R_j}\frac{d\omega_j}{dt} > \frac{d\gamma}{dt} \tag{4.17}$$

Substituting (4.17) into (4.16) yields

$$\left(I + mR^2 \right) \frac{r}{R_j}\frac{d\omega_j}{dt} > \sum_{i=1}^{4} \left[\left(\tau_i - J\,\frac{d\omega_i}{dt} \right) \frac{R_i}{r} \right] + RF_R \tag{4.18}$$

$$\frac{d\omega_j}{dt} > \frac{\sum\limits_{\substack{i=1 \\ i \neq j}}^{4} \left[R_i \left(\tau_i - J\, d\omega_i/dt \right) \right] + r\,RF_R + \tau_j R_j}{\left(I + mR^2 \right) r^2/R_j + J\,R_j} \tag{4.19}$$

As F_R is always negative, wheel j is definitely slip when the following inequality is satisfied:

$$\frac{d\omega_j}{dt} > \frac{\sum\limits_{\substack{i=1 \\ i \neq j}}^{4} \left[R_i \left(\tau_i - J\, d\omega_i/dt \right) \right] + \tau_j R_j}{\left(I + mR^2 \right) r^2/R_j + J\,R_j} \tag{4.20}$$

The proposed wheel-slip observer (4.20) only requires the data of the torque and acceleration of all wheels, which are easy to obtain.

When the steering angle tends to zero, R_i, R_j, and R tend to infinity. Then (4.20) becomes

$$\frac{d\omega_j}{dt} > \frac{\sum\limits_{\substack{i=1 \\ i\neq j}}^{4} (\tau_i - J\, d\omega_i/dt) + \tau_j}{J + mr^2} \tag{4.21}$$

When only one wheel is considered, (4.21) becomes

$$\frac{d\omega}{dt} > \frac{\tau}{J + mr^2} \tag{4.22}$$

which is the same as the wheel-slip observer proposed in Fujimoto, Tsumasaka, and Noguchi.[25] It shows the generality of the proposed wheel-slip observer.

4.2.1.2 Yaw Rate Error Observer

According to the spinning speed of a wheel ω_i and the target ICR, the desired yaw rate γ_t of a vehicle body is defined as

$$\gamma_t = \frac{\omega_i r}{R_i} \tag{4.23}$$

Obtaining γ_t according to a different wheel will result in a slightly different value because of the different amount of slip on different wheels. To moderate the error, γ_t is redefined as the minimum value of the solutions obtained according to the spinning speeds of different wheels:

$$\gamma_t = \min\left(\frac{\omega_i r}{R_i}\right) \qquad \forall i \in \{1, 2, 3, 4\} \tag{4.24}$$

The actual yaw rate γ_m of the vehicle body can be measured by a gyroscope. The yaw rate error is then defined as

$$\gamma_e = \gamma_t - \gamma_m = f(\phi, \omega_i, \gamma_m) \tag{4.25}$$

As wheel slip will make γ_t an overestimate, wheel slip must be suppressed before applying the DYC.

4.2.1.3 Heading-Error Observer

In natural steering, the centripetal acceleration $\mathbf{a}_c$ is perpendicular to the target heading direction β_t. If a vehicle is not in natural steering, a heading error β_e exists. The measurement of a two-axis accelerometer $\mathbf{a}_m$ includes the actual tangential acceleration ${}^a\mathbf{a}_t$ and centripetal acceleration ${}^a\mathbf{a}_c$:

$$\mathbf{a}_m = {}^a\mathbf{a}_t + {}^a\mathbf{a}_c = \begin{bmatrix} a_x & a_y \end{bmatrix} \tag{4.26}$$

where a_x and a_y are the orthogonal readings in the two-axis accelerometer. Let the measurement of a_x and a_y be along x and y axes as shown in Fig. 4.12. The relationship between the readings and the actual state can be written as

$$\begin{bmatrix} {}^a a_t \\ {}^a a_c \end{bmatrix} = \begin{bmatrix} \cos(\beta_e + \beta_t) & \sin(\beta_e + \beta_t) \\ -\sin(\beta_e + \beta_t) & \cos(\beta_e + \beta_t) \end{bmatrix} \begin{bmatrix} a_x \\ a_y \end{bmatrix} \tag{4.27}$$

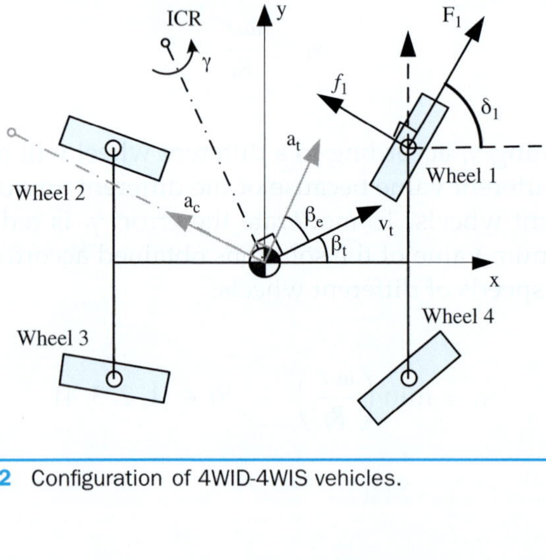

FIGURE 4.12 Configuration of 4WID-4WIS vehicles.

It is assumed that the vehicle velocity at the target heading direction can be approximated by the spinning speed of a wheel by

$$v_t = \gamma_t R \tag{4.28}$$

where γ_t is obtained from (4.24).

The linear acceleration of the vehicle at target heading direction a_t can be obtained from the derivative of v_t

$$a_t = \frac{d}{dt} v_t = R \frac{d}{dt} \gamma_t \tag{4.29}$$

As the centripetal acceleration cannot be observed through wheel spinning speeds, it can be concluded that

$${}^{a}a_t \cos \beta_e = a_t \tag{4.30}$$

Substituting (4.30) into (4.27) gives

$$\left[a_x \cos(\beta_e + \beta_t) + a_y \sin(\beta_e + \beta_t)\right] \cos \beta_e = a_t \tag{4.31}$$

The heading error can then be obtained by solving (4.31), represented as

$$\beta_e = f(\phi, \omega_i, a_m) \tag{4.32}$$

As (4.31) is a nonlinear equation, β_e can be obtained by solving the equation using an iteration method such as Newton's method. In (4.31), a_x, a_y, and a_t are measurable by a two-axis accelerometer and wheel speed sensors. No time-varying parameter or integration process is needed in the calculation; hence this makes the heading-error observer accurate and practical.

As the observer is based on centripetal force detection, it can approximate the heading error only in turning motion. The heading error in pure translation motion cannot be measured, whereas the heading error will rarely occur in pure translation motion in the horizontal plane. If a vehicle moves almost in pure translation motion, a small amount of measurement error will induce a great amount of error of the heading error approximation. To eliminate the error, when the detected centripetal force is smaller than a threshold ε

$$^{a}a_c = -\sin(\beta_e + \beta_t)\, a_x + \cos(\beta_e + \beta_t)\, a_y < \varepsilon \tag{4.33}$$

Here β_e is set as zero as it is assumed that the heading error will not occur in pure translation motion.

4.2.2 Feedforward and Feedback Control

4.2.2.1 Feedforward Control

To suppress the side slip of wheels, the feedforward controller calculates the required yaw moment to turn the vehicle in turning motion according to the steering inputs and linear acceleration at the target heading direction a_t

$$^{f}M = I\frac{a_t}{R} = f(\phi, \omega_i) \tag{4.34}$$

Then the feedforward controller generates it by longitudinal traction forces so as to reduce the lateral force exerted on the wheels.

4.2.2.2 Feedback Control

Generally, there is a limitation in using the feedforward controller only to control a vehicle, especially when the vehicle runs into some unexpected changes such as the change of the condition of road surfaces. Therefore, it is necessary to additionally provide a feedback controller to compensate for such disturbance.

In the feedback controller, a clockwise yaw moment is generated when $\gamma_e > 0$ to compensate for the error. When $\gamma_e < 0$, an anticlockwise yaw moment is generated. This control scheme makes the yaw rate of the vehicle follow the desired yaw rate and hence improves vehicle steerability. However, it also increases the heading error of the vehicle. It is reported that most traffic accidents occur due to excessive heading error of vehicles.[40] Therefore, it is important to suppress the heading error of vehicles. If β_e is large, a serious side slip occurs. The vehicle should adjust the orientation to obtain traction. Hence, when $\beta_e > 0$, a clockwise yaw moment is required.

Similarly, when $\beta_e < 0$, an anticlockwise yaw moment is required.

With the control inputs β_e and γ_e, numerous control techniques can be used, such as PID control or fuzzy logic control, to suppress β_e and γ_e. Take proportional gain control as an example. The feedback yaw moment to be generated becomes

$$^{b}M = K_\beta \beta_e + K_\gamma \gamma_e \tag{4.35}$$

where K_γ and K_β are the positive proportional gains. Since β_e has first priority to be suppressed, K_β must be selected greater than K_γ.

By combining the feedforward and feedback controller, the total yaw moment to be generated becomes

$$M_{\text{req}} = {}^{f}M + {}^{b}M = f(\phi, \omega_i, \gamma_m, a_m) \tag{4.36}$$

It can be realized that the controller takes only the steering inputs ϕ, wheel spinning speeds ω_i, yaw rate of the vehicle body γ_m, and linear acceleration of the vehicle body a_m as control inputs, which can be obtained easily by conventional sensors, such as the wheel speed sensor, gyroscope, and accelerometer. The assumption is that the wheels have no serious slip in the longitudinal direction, which can be detected by the proposed wheel-slip observer.

In Sec. 4.1 we have developed a deterministic force generation algorithm to generate desired yaw moment and lateral force deterministically by reallocating the force distribution on each wheel. The merit over the traditional method which differentially distributes the force on left and right wheels is that it considers the nonlinearity of the turning radius and generates the exact yaw moment in any ICR. According to Sec. 4.1, the traction force on wheel i (F_i) is defined as

$$F_i = O_i + N_i \varepsilon_n + M_i \varepsilon_m \tag{4.37}$$

where O_i, N_i, and M_i are functions of the position of wheel i and the steering command ϕ; ε_n and ε_m are the centripetal force and yaw moment to be generated by longitudinal traction forces, respectively. As a result, in the controller, ε_n and ε_m are set as

$$\varepsilon_m = M_{\text{req}} \qquad \varepsilon_n = 0 \tag{4.38}$$

Parameter	Symbol	Unit	Value
Vehicle total mass	m	kg	1000
Moments of inertia of vehicle	I_z	$kg \cdot m^2$	1600
Coordinates of wheel 1	(x_1, y_1)	m	(1.25,0.75)
Coordinates of wheel 2	(x_2, y_2)	m	(−1.25,0.75)
Coordinates of wheel 3	(x_3, y_3)	m	(−1.25,−0.75)
Coordinates of wheel 4	(x_4, y_4)	m	(1.25,−0.75)

TABLE 4.4 Parameters of the Vehicle Model

4.2.3 Performance Analysis

Numerous simulation studies have been conducted to evaluate the performance of the heading-error observer and the proposed controller before the experiments on real vehicles are carried out. The parameters of the vehicle model used in the simulation are shown in Table 4.4. The yaw moment controller adopts the proportional gain control method as described in (4.36). Uniform white noise with amplitude 0.01 is added in the measuring inputs such as the accelerometer and wheel speed readings to imitate the actual situation in real-world measurement.

4.2.3.1 Performance of the Heading-Error Observer

To evaluate the performance of the proposed heading-errors observer, the actual and estimated heading errors in different road conditions and steering inputs have been evaluated.

In the first set of experiments, different road conditions, as illustrated in Fig. 4.13, have been tested with a step steering command. The actual and estimated heading errors in different road conditions are shown in Fig. 4.14. In the figures, the black and blue lines represent the actual and estimated heading error, respectively. It shows that without using the knowledge of the dynamic parameters between tires and the road surface, the observer can estimate the heading error with high accuracy. The estimation error is mainly due to the noise of measurement, which can be moderated by noise filtering techniques. It shows that the proposed observer is robust and practical and has high potential to be applied on commercial vehicles.

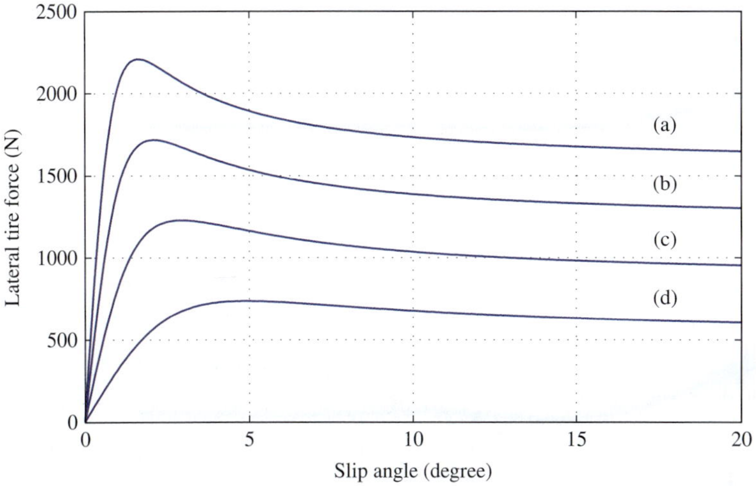

FIGURE 4.13 Four different road conditions tested in the experiments.

In the second set of experiments, the vehicle was commanded in different steering inputs, i.e., single lane change, step, and sinusoidal inputs. The road condition (a) as illustrated in Fig. 4.13 is adopted. The actual and estimated heading errors in different steering inputs are illustrated in Fig. 4.15. It can be observed that the estimated heading errors are close to the actual values in different steering commands. It can also be observed that when the heading error is close to zero, the variation of the estimation becomes larger because at that time the vehicle was almost in pure translation motion. It shows that the estimated heading error is sensitive to the noise of measurement in this state which matches the expectation. The serious estimation errors were then suppressed to zero when a small centripetal force was detected.

4.2.3.2 Performance of the DYC

To evaluate the performance of the proposed DYC, a single-lane-change maneuver was commanded as shown in Fig. 4.16. In the simulation, a front-wheel steering configuration was adopted. The initial vehicle speed was set as 30 m/s. The trajectories of the vehicle using the proposed DYC and the zero yaw moment method as proposed in Sec. 4.1 are illustrated

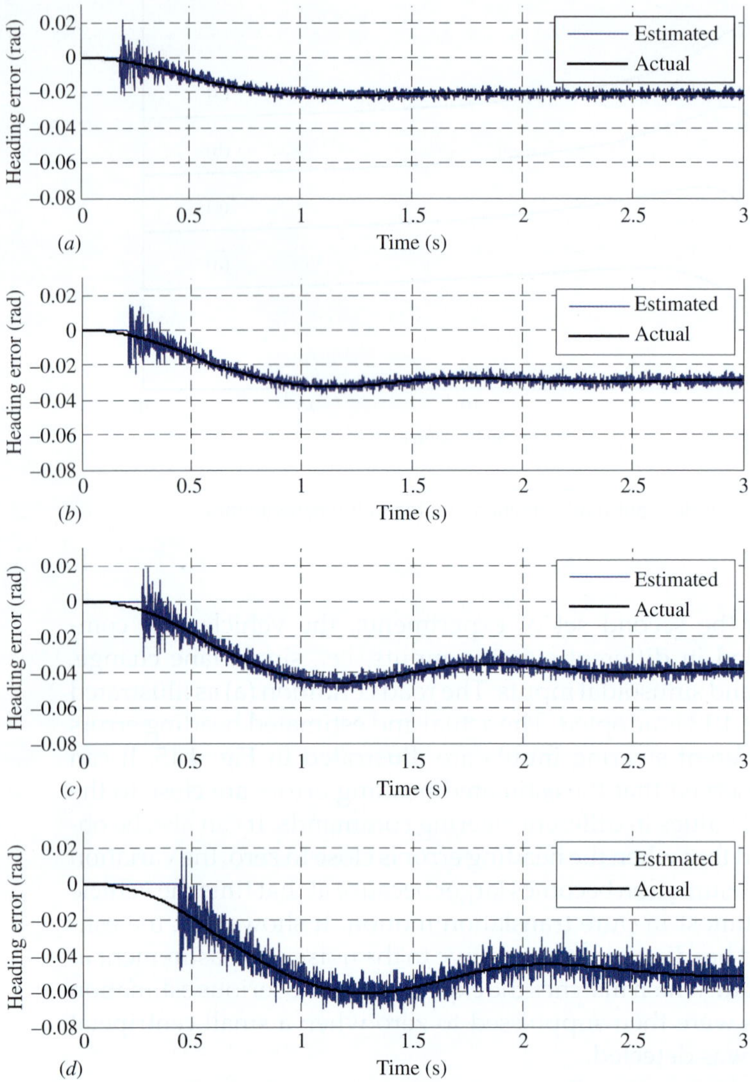

FIGURE 4.14 Actual and estimated heading errors in different road conditions.

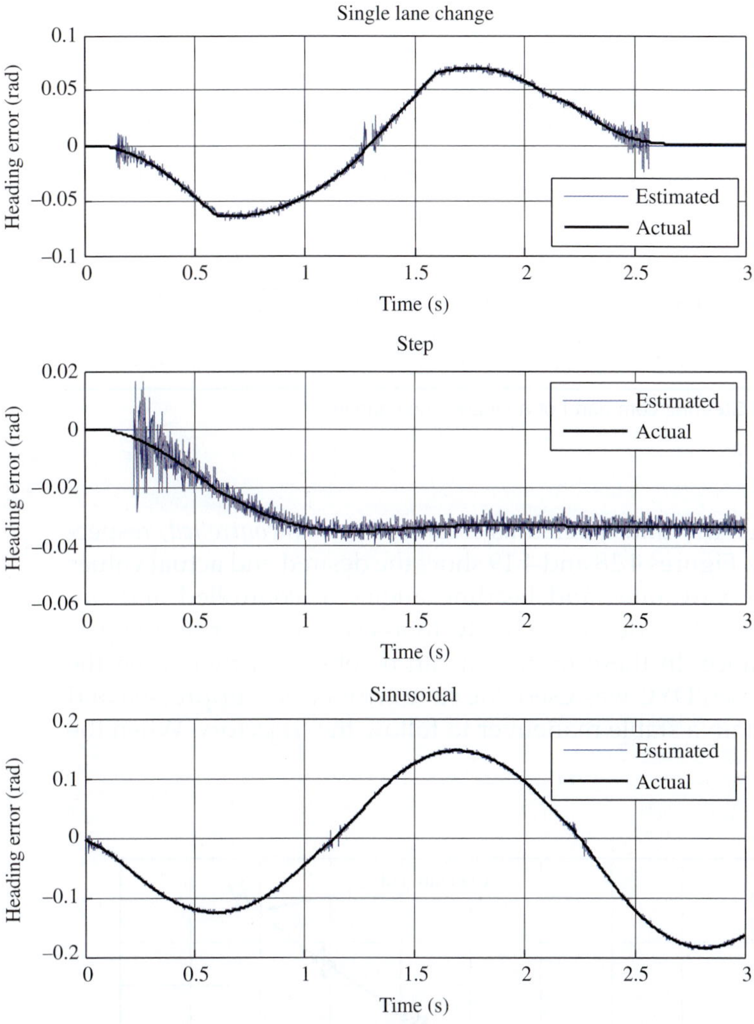

FIGURE 4.15 Actual and estimated heading errors by the proposed heading-error observer in different steering commands.

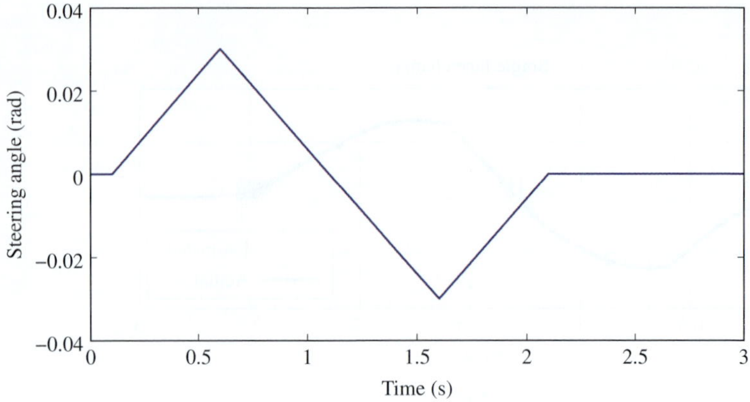

FIGURE 4.16 Steering command of a single lane change.

in Fig. 4.17 and labeled as *controlled* and *uncontrolled*, respectively. Figures 4.18 and 4.19 show the desired and actual values of the yaw rates and heading angles in controlled and uncontrolled mode, respectively, to assist in comparing the performance. In these figures, it can be observed that when the proposed DYC was used, the state errors were suppressed and maintain a stable maneuver to follow the trajectory. When the

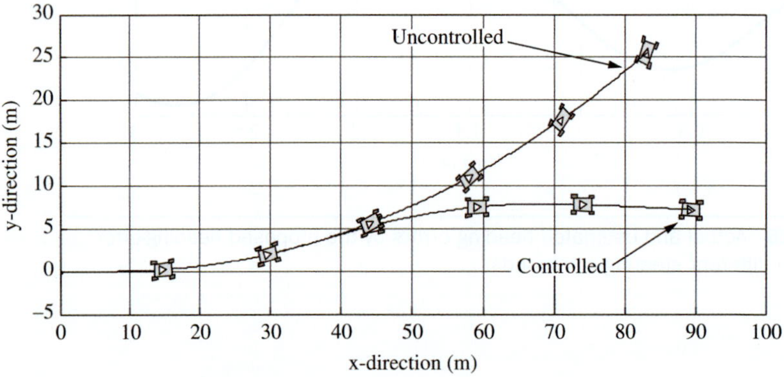

FIGURE 4.17 Trajectories of the sudden single-lane-change maneuver with and without using the proposed DYC.

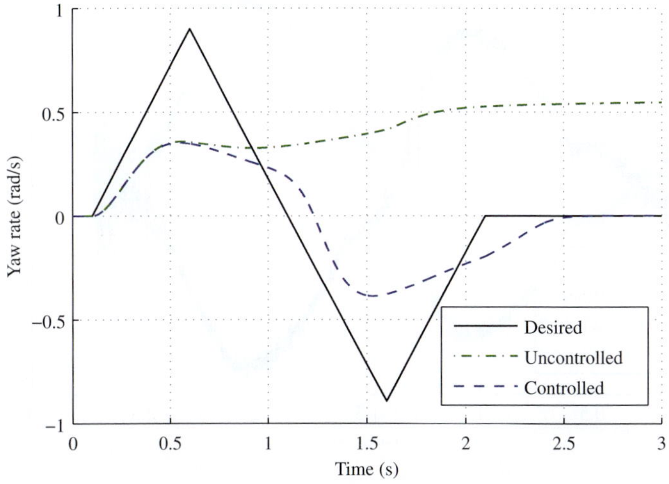

FIGURE 4.18 Desired and actual yaw rates with and without using the proposed DYC.

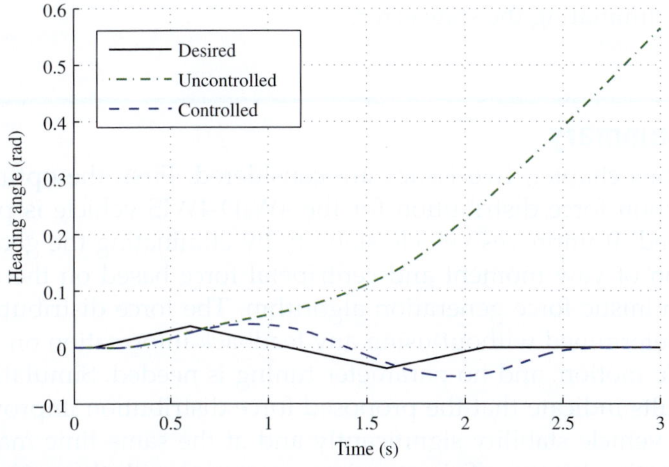

FIGURE 4.19 Desired and actual heading angles with and without using the proposed DYC.

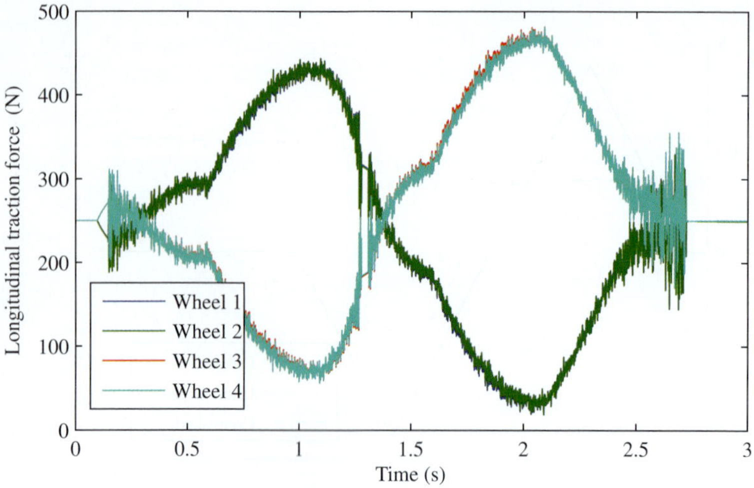

FIGURE 4.20 Longitudinal traction force distribution using the proposed DYC.

DYC is disabled, the state errors increase continuously and make the vehicle fail to follow the desired trajectory. The longitudinal traction force distributions on wheels by using the proposed DYC are shown in Fig. 4.20. Each wheel has a different force assignment to generate the necessary yaw moment for eliminating the state errors.

4.3 Summary

In this chapter, two issues are considered. First, the optimal traction force distribution for the 4WID-4WIS vehicle is proposed. It improves vehicle stability by eliminating the generation of yaw moment and centripetal force based on the deterministic force generation algorithm. The force distribution is determined without using any feedback information on vehicle motion, and no parameter tuning is needed. Simulation results indicate that the proposed force distribution improves the vehicle stability significantly and at the same time maintains the driving efficiency when compared with the uniform force distribution.

Second, a general wheel-slip observer that considers all wheels and omnidirectional steering is introduced. It can be applied on vehicles with any steering configuration. In addition, a heading-error observer that is independent of the tire dynamic and the characteristics of the road surface is proposed. Simulation results reveal the accuracy and robustness of the proposed observer in different road and steering conditions. Moreover, a direct yaw moment controller for a 4WID-4WIS vehicle based on the proposed observers and the deterministic force generation method is presented. Simulations have been conducted to evaluate the performance, and the results are satisfactory.

CHAPTER 5

Battery Management System

This chapter discusses the battery management system (BMS) for HEV. Section 5.1 illustrates the system structure and hardware design of the BMS prototype. Section 5.2 focuses on the robust state of charge (SOC) estimation framework, including an adaptive-κ nonlinear diffusion filter to reduce noise and a self-learning mechanism to remove zero drift in the measurement of current. Then two models, the resistive-capacitive (*RC*) model and modified enhanced self-correcting (ESC) model, are considered for simulation and experiment, respectively. Section 5.3 proposes a fast-charge framework based on model predictive control (MPC), with the aim of simultaneously reducing charge duration, which represents safety and energy efficiency during the charge process. Section 5.4 proposes a SOC-based fuzzy controller to equalize cells. Both the simulations and experiments are designed to test the performance of the proposed methods.

5.1 Hardware Design

A vehicular battery pack must consist of a large number of cells to provide the necessary energy and power to drive EVs. Unlike with portable devices, it is impossible to manage the vehicular battery only at the pack level, which leads to two problems: out-of-investigation cells and lack of cell equalization ability.

First, the state of cells cannot simply be calculated according to the state of the pack. Cell imbalance causes the state of some cells to be far away from the average, especially when there are a large number of cells. These abnormal cells, which should be closely supervised, are out of the range of investigation and hence risk abuse and even damage, such as overcharging, overdischarging, and so forth. Furthermore, if one cell is broken, the pack manager cannot identify it; the whole pack has to be replaced and then repaired off-line.

Second, management at pack level cannot equalize the cells. This disadvantage not only intensifies the first problem, but also limits the performance of the whole pack. A battery is a typical system with a cask effect, in which the weakest cell determines the discharge ability of the pack, while the strongest determines its charge ability. For example, given the extreme case in which one cell's SOC is 1 while another's is 0, how should the battery pack operate? Charging will cause overcharge damage to the full cell, while discharging cannot avoid overdischarge damage to the empty one. Without cell equalization, the battery pack is unserviceable even though the pack still contains a large number of charges.

The earlier analysis reveals that a BMS for vehicular batteries must have a specially designed system structure and must manage the pack as well as the individual cells.

The first option is a centralized structure. In this structure, the pack and all cells are managed by a single microprogrammed control unit (MCU).[19, 88] Due to the large number of cells, the MCU has to measure and control cells in turn, using selective circuits. Obviously, this design cannot investigate cells in real time. In addition, it is very hard to estimate the SOC of each cell in a single MCU.

A hierarchical structure (also known as a modular BMS) is therefore a more reasonable design, in which first cells are grouped into modules managed by modular controllers and then modules are incorporated into a pack managed by a central controller. Design considerations for a modular BMS include the following:

1. How to distribute tasks to modular and central controllers
2. How to communicate with one another
3. How to measure and control cells and the pack
4. How to provide power supply for controllers

5.1.1 System Structure and Function Analysis

The system structure of the BMS in this book follows the modular concept and its implementation proposed in Ref. 102; however, it includes some modifications to improve accuracy, modularization, and intelligence. The following sections introduce the structure and function of each component of the smart module and central controller one by one.

5.1.1.1 Smart Module

A schematic diagram of a smart module is shown in Fig. 5.1. A smart module applies two AD7280 ICs to measure the voltages of 12 cells. Four temperature sensors are fixed in the module to roughly measure temperature. All the data are sent to the module controller MC9S12XEP100MAL.[24] SOC estimation, overcharge and overdischarge detection, and equalization control are all implemented in the controller in a soft manner.

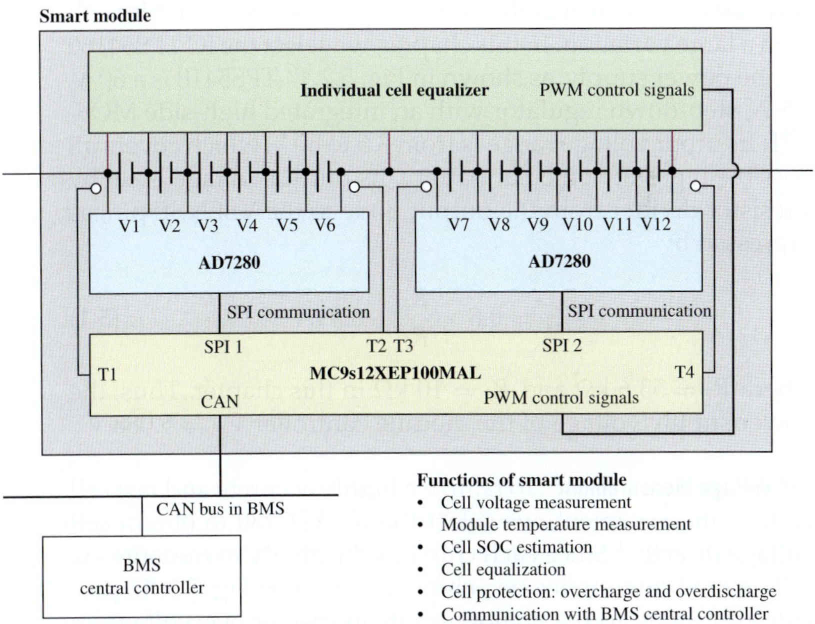

FIGURE 5.1 Structure and functions of smart module.

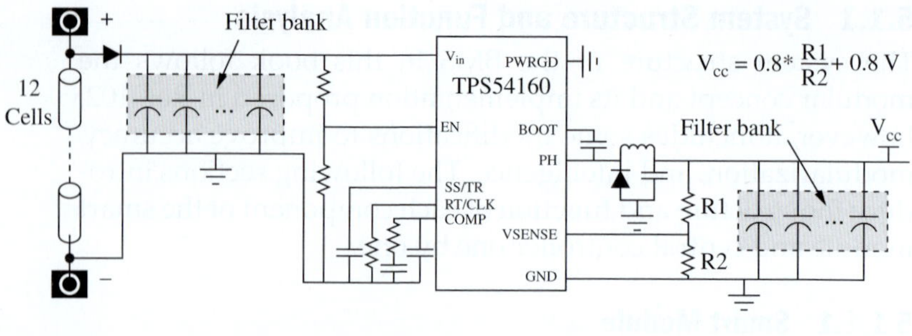

FIGURE 5.2 Schematic diagram of power supply for module controllers.

Power Supply The cells inside the module provide the power source for the module controller. Since each module contains 12 cells, the voltage of the power supply may vary from 30 V (2.5-V cutoff voltage) to 50.4 V (4.2-V full voltage) for a Li-ion battery, and from 12 V (1-V cutoff voltage) to 16.8 V (1.4-V cutoff voltage) for a NiMH battery. To develop a universal BMS, it is necessary to ensure that the power supply is always stable with such a large variation. In this chapter, we select the IC TPS54160 for the power supply, as shown in Fig. 5.2.[107] TPS5410 is a 60-V, 1.5-A, step-down regulator with an integrated high-side MOSFET. Its input voltage range is from 3.5 to 60 V, which covers all possible types of vehicular battery. Its output voltage is set by a resistor divider from the output node to the VSENSE pin, as expressed by

$$V_{cc} = 0.8 \times \frac{R_1}{R_2} + 0.8V \tag{5.1}$$

where $R_1 = 53.6\ \text{k}\Omega$ and $R_2 = 10\ \text{k}\Omega$ in this chapter. Thus, the power supply voltage of the module controller $V_{cc} = 5.088$ V.

Cell Voltage Measurement To realize a highly accurate and fast cell voltage measurement, we select the IC AD7280 to obtain cell voltage directly.[1] Since each chip has the ability to measure six cells, a module contains two chips, as shown in Fig. 5.3. To provide protection to the analog inputs in case of overvoltage or undervoltage of cells, for each channel a 10-kΩ resistor is connected with a cell in series. In conjunction with the resistor, a 100-nF capacitor across the pseudo differential inputs acts as a

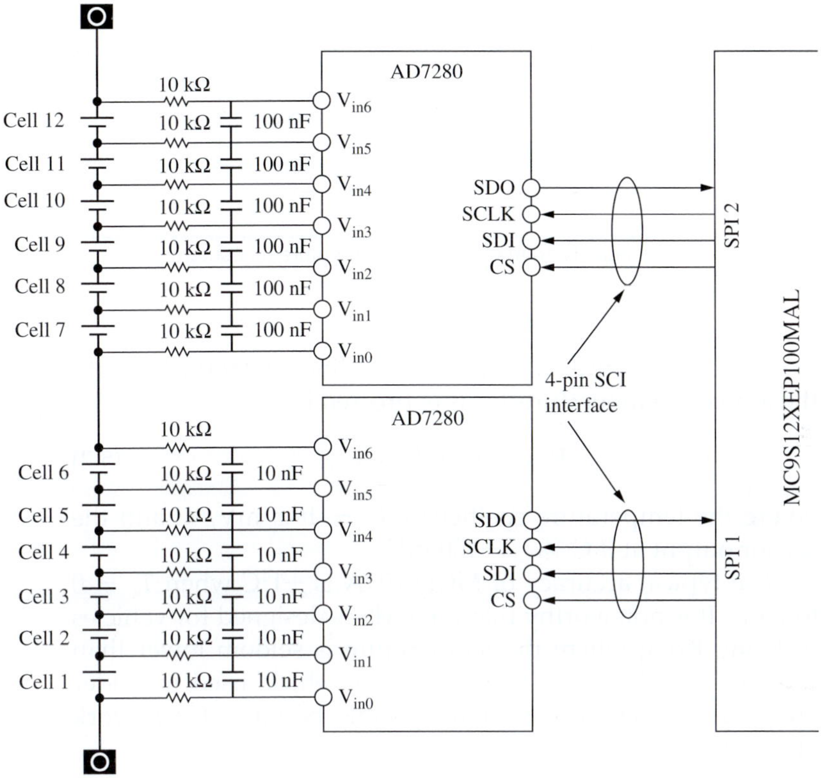

FIGURE 5.3 Schematic diagram of voltage measurement.

low-pass filter, with a cutoff frequency of 318 Hz. This external measurement configuration allows a combined acquisition and conversion time of 1 μs, which is fast enough to be recognized as real-time measurement. To send the voltage measurement results, the two AD7280s both communicate with the MCU via SPI ports, rather than via daisy chain. Its typical accuracy is 0.07 percent (−40 to 85°C), equal to 2.8 mV for a 4-V cell (1.4 percent SOC). It has the best performance for voltage measurement, as discussed in the previous section.

Temperature Measurement We apply the IC MCP9701A to measure the temperature inside the module.[63] MCP9701A is the most accurate of the low-power linear active thermistors in the MCP9000 family. It has a very simple external circuit, as shown

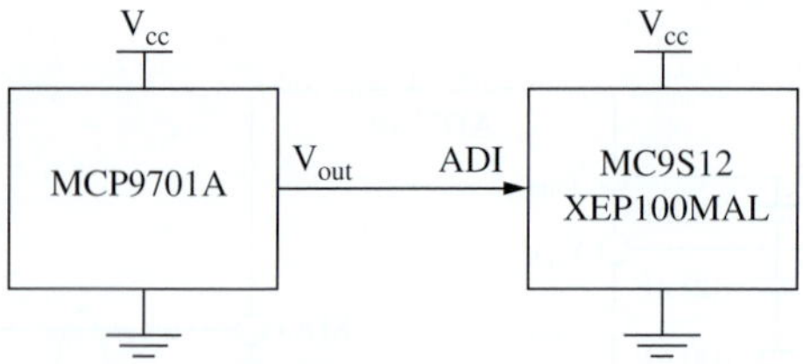

FIGURE 5.4 Schematic diagram of module temperature measurement.

in Fig. 5.4. The output voltage V_{out} has a linear relationship with the ambient temperature T_a, as expressed by

$$V_{out} = T_c \times T_a + V_{0^\circ C} \tag{5.2}$$

where the temperature coefficient $T_c = 19.5$ mV/°C and the sensor output at 0°C $V_{0^\circ C} = 400$ mV.

The typical accuracy of MCP9701A is ±1°C when $T_a = 0$ to 70°C. It is noteworthy that the BMS is designed for vehicles in Hong Kong where the temperature is seldom lower than 0°C. If the BMS is equipped within a vehicle used in colder areas, the MCP9700A is more suitable because it can work at −25°C.

Cell Equalization Cell equalization is implemented in two ways in this chapter.

The first method is the traditional discharging equalization, which discharges the cells with higher SOC through dissipative resistors. With the advantage that the interface for cell equalization is already integrated in the AD7280, the implementation of discharging equalization is quite simple, as shown in Fig. 5.5.

When the MCU detects that one cell's SOC is higher than others, e.g., if it exceeds a threshold, the MCU will send the equalization command to the AD7280, which will make the corresponding MOSFET discharge the cell through its bypass resistor. The resistances of R_1 to R_{12} determine the discharging current of cells in equalization. Selection of the resistance should take two points into consideration. For discharging equalization, the equalizing current cannot be too large, typically 0.01 C. Meanwhile, the discharging power cannot exceed the bearing capacity of the circuit board. In this

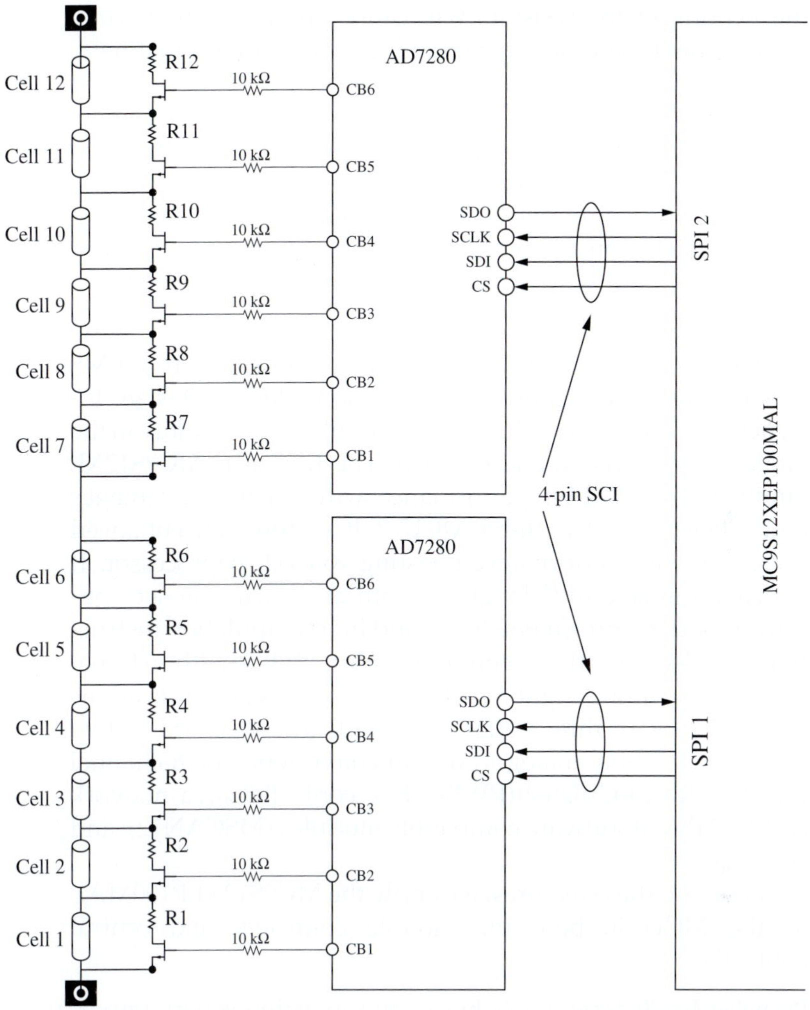

FIGURE 5.5 Schematic diagram of cell equalization.

work, the resistances are set to 46 Ω, corresponding to about 0.069-A equalizing current for the 6.9-Ah LiFePO4 cells in the application.

Although equalizing cells using this discharging method are quite simple, and it has been applied in many existing EVs, some disadvantages hinder its future promotion. First, it is a dissipative method causing energy loss on the resistors. Second, the heating of the resistors will cause problems for thermal management, especially for retaining a uniform thermal field. In addition, the discharging current is a constant determined by the discharging resistor and is usually designed to be small. This feature causes a long equalization period, especially when the imbalance is excessive.

Therefore, to improve the quality of equalization, it is necessary to develop an advanced equalization method. In Chap. 4, an intelligent transferring equalization approach has been proposed to study this issue further.

Microprogrammed Control Unit Selection of the MCU in a BMS should consider cost, power consumption, EMC, code-size efficiency, communication ability, adaptability and reliability in the vehicle environment, and so forth. The free-scale MC9S12XE family delivers 32-bit performance with all the advantages and efficiencies of a 16-bit MCU.[24] It features an enhanced version of the performance-boosting XGATE coprocessor. It is programmable in C language with an instruction set optimized for data movement, logic, and bit manipulation instructions. It also has strong communication ability with external devices, including eight synchronous serial communications interfaces (SCI); three serial peripheral interfaces (SPI); two 16-channel, 12-bit analog-to-digital converters; one 8-channel pulse-width modulator (PWM); five controller area network (CAN) 2.0 A, B software compatible modules (MSCAN12); and so on.

Based on these features, we apply the MC9S12XEP100MAL as the MCU in both the module controller and central controller.

Controller Area Network CAN bus communication is very important for automotive engineering. It is realized in two steps.

First, ADuM5402 is applied to isolate the signals in the controllers from the CAN bus, as shown in Fig. 5.6. The ADuM5402 isolator provides four independent isolation channels in a variety of channel configurations and data rates. The number of channels is sufficient to use a single chip to handle the isolation of CAN communication.

Next, the high-speed CAN-transceiver TLE6250 is applied to work as an interface between the CAN protocol controller

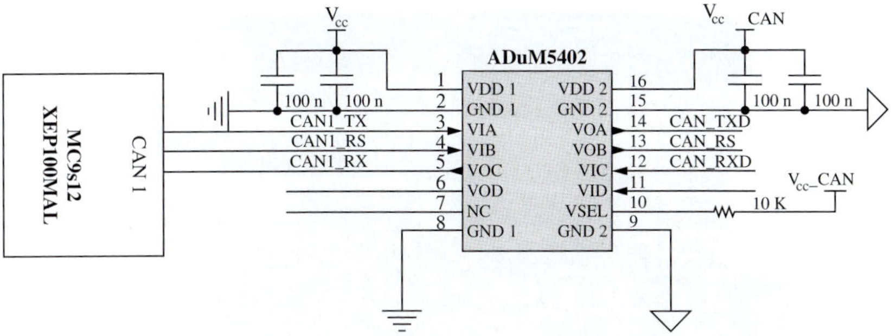

FIGURE 5.6 Schematic diagram of CAN isolation.

and the physical differential bus, as shown in Fig. 5.7.[39] TLE6250 is optimized for high-speed differential mode data transmission in automotive and industrial applications. Protection of electrostatic discharge (ESD) is implemented based on the dual-line CAN bus protector NUP2105L.[73] To improve anti-EMI ability, the B82793, the choke for data and signal lines, is also series-connected with the CAN lines.[21]

Implementation Figure 5.8 shows the photograph of the implemented module controller. It is 95 mm × 95 mm × 11.6 mm, with four layers. The small size allows it to be fixed in a small volume.

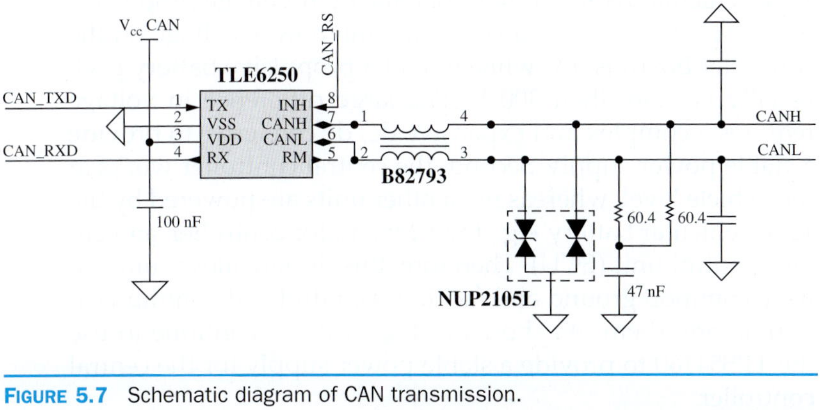

FIGURE 5.7 Schematic diagram of CAN transmission.

FIGURE 5.8 Photograph of the implemented module controller.

5.1.1.2 Central Controller

The schematic diagram of the central controller circuits is shown in Fig. 5.9. It mainly focuses on management at the pack level, including measurement of pack voltage, pack current, and current leakage of the pack. It also drives fans and contactors to control pack temperature and the opening and closing of the battery power bus.

Power Supply The power supply of the central controller is the 12-V vehicular battery rather than the high-voltage propulsive battery pack, for two reasons. First, the power voltage on the controller board is 5 V, while that of a propulsive battery pack usually is more than 300 V. The large difference in voltage requires a complex and expensive dc/dc converter to provide a stable power supply. Second, the central controller works at the vehicle level, whereas most other units are powered by the 12-V vehicular battery, e.g., the VMS, motor controller, and engine control unit (ECU). Therefore, this design allows units to have common ground and facilitates control and communication among them. As shown in Fig. 5.10, we continue to use the TPS54160 to provide a stable power supply for the central controller.

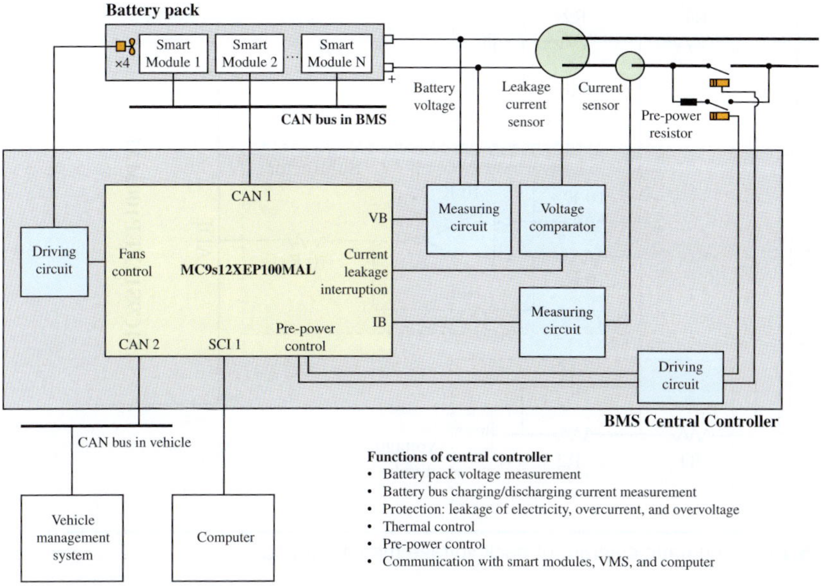

FIGURE 5.9 Structure and functions of central controller.

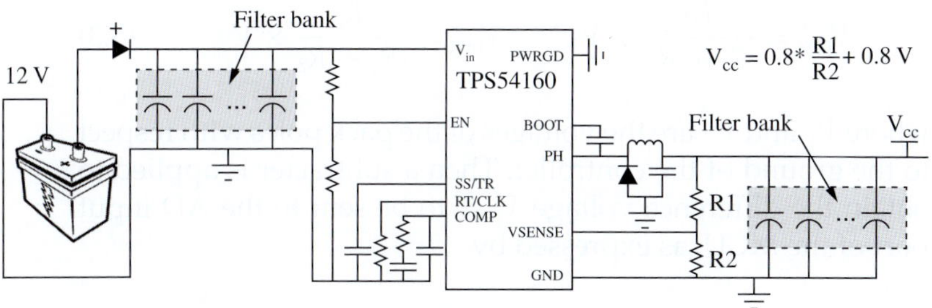

FIGURE 5.10 Schematic diagram of power supply for the central controller.

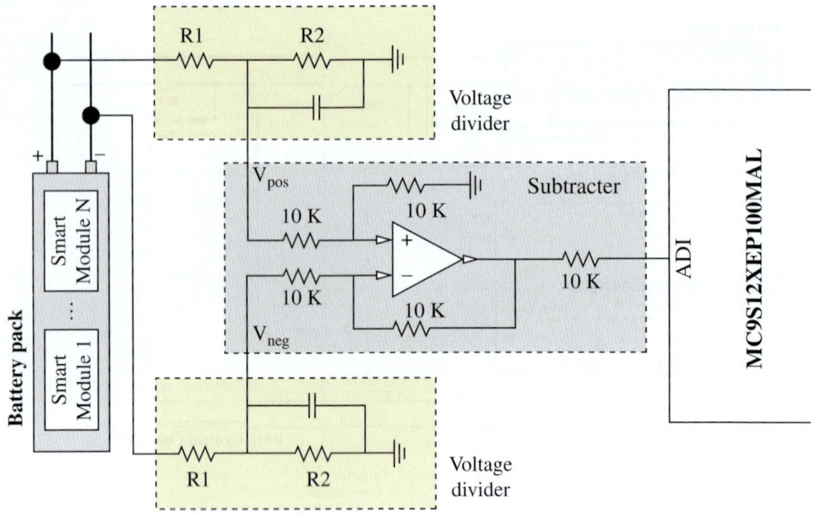

FIGURE 5.11 Schematic diagram of pack voltage measurement.

Battery Pack Voltage Measurement As shown in Fig. 5.11, to measure the high pack voltage, the first step is to use voltage dividers to linearly reduce the voltages of positive and negative battery poles below 5 V. It is noteworthy that the negative pole of the battery pack is not the ground of the central controller. The two low voltages V_{pos} and V_{neg} are respectively calculated as

$$V_{pos} = \frac{R_2}{R_1 + R_2} \times V_+ \qquad V_{neg} = \frac{R_2}{R_1 + R_2} \times V_- \tag{5.3}$$

where V_+ and V_- are the voltages of the pack poles with respect to the ground of the controller. Then a subtracter is applied to obtain the difference voltage V_{pack} to be sent to the AD input port of the MCU, as expressed by

$$V_{pack} = V_{pos} - V_{neg} = \frac{R_2}{R_1 + R_2} \times (V_+ - V_-) \tag{5.4}$$

The selection of R_1 and R_2 depends on the voltage of the battery pack, making sure V_{pack} is below 5 V to avoid damage to the MCU.

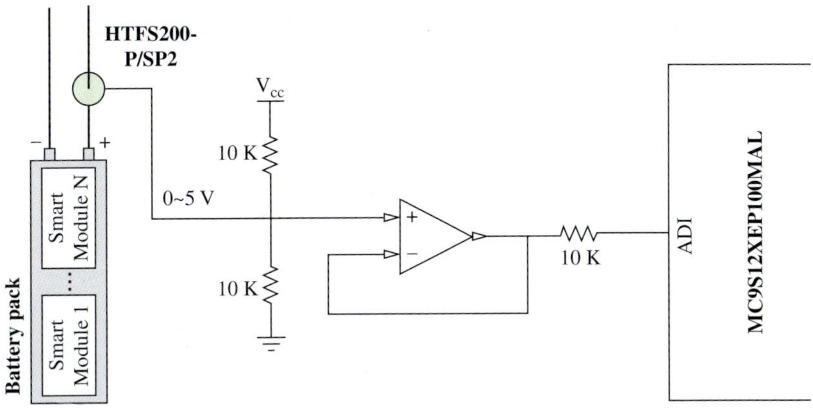

FIGURE 5.12 Schematic diagram of pack current measurement.

Battery Pack Current Measurement The Hall effect current sensor HTFS200-P/SP2 [56] is applied to convert the current in ±300 A to the voltage in 2.5 ± 1.875 V, as expressed by

$$V_{out} = 2.5 + 1.25 \times \frac{I}{200} \text{ V} \tag{5.5}$$

where I is the pack current to be measured and V_{out} is the output voltage.

To decrease output impedance and isolate the central controller from the sensor fixed on the power bus, V_{out} is sent to the AD input port via a follower, as shown in Fig. 5.12.

Leakage Current Measurement When current leakage occurs, a small amount of current goes to the vehicle body or the ground, making the current on the battery's positive pole different from that on the negative pole. Measurement of the small current difference allows detection of current leakage.

The CYCT04-M20B current sensor family is based on a magnetic modulation and compensation principle, and it can be used for measurement of small dc current and leakage current, a current difference between two or more conductors.[12] Leakage current is bidirectional. The sensor linearly converts the leakage current in ±50 mA to the output voltage ±5 V.

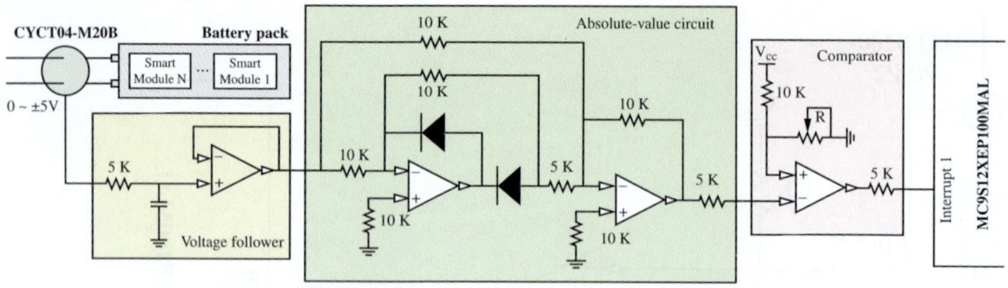

FIGURE 5.13 Schematic diagram of leakage current measurement.

As shown in Fig. 5.13, the output voltage of the sensor first goes through a voltage follower to decrease output impedance and isolate the central controller from the sensor. Then an absolute-value circuit is applied to convert the voltage in ±5 V to 0 to 5 V. This is so because the current leakage is related to only the value, but not the direction, of the leakage current. Finally, a comparator is designed to produce an interruption signal if the measured voltage is higher than the reference voltage determined by the sliding resistor *R*. In this work, *R* is selected to produce interruption if the value of the leakage current is greater than 30 mA.

Driving Circuit of Fans and Contactors Fans and contactors are applied to control the battery package temperature and the on/off of the power bus, respectively. To drive fans and contactors, the driving circuit shown in Fig. 5.14 is applied for each channel. The NCV33125, the dual high-speed MOSFET driver, is applied to drive an actuator via a MOSFET.[72] Since the power of fans and contactors is higher than the driving ability of the controller, we directly use the 12-V vehicle battery for the power source. In addition, to measure the states of fans and contactors, each channel is also fed back to the MCU.

Implementation Figure 5.15 shows the photograph of the implemented central controller. It is 120 mm × 95 mm × 11.6 mm, with four layers; as it is the same width and height as those of the modules, assembly is easy.

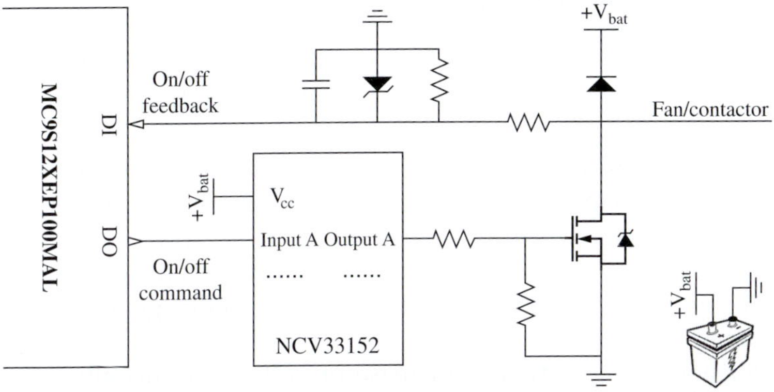

FIGURE 5.14 Schematic diagram of driving circuit of fans and contactors.

FIGURE 5.15 Photograph of the implemented central controller.

5.1.1.3 Battery Pack with Battery Management System

To validate the feasibility of the proposed BMS and establish a platform to study the proposed SOC estimation, cell equalization, and fast charging algorithms in the following chapters, a battery pack with the BMS has been developed.

The pack consists of 2.3-Ah LiFePO4 cells, which have ordinary performance including charge and discharge ability, internal resistance, cell consistency, and so forth. The aim of selecting this pack is to demonstrate the efficiency of the proposed BMS and various algorithms under general conditions.

Extension of capacity and increase of voltage are usually two necessary requirements for establishing a vehicular battery. Extension of capacity is implemented by connecting cells in parallel, while increase of voltage is implemented by connecting cells in series. In this case, two pack architectures are possible. The first approach, called a series cell module, builds the pack by first wiring the cells in series for increased voltage and then connecting the high-voltage strings in parallel for extension of capacity. In contrast, the second method, called a parallel cell module, creates the pack by first wiring cells in parallel and then connecting the large-capacity strings in series. We follow the second approach in this work, with the advantage that the parallel-connected cells can be recognized as a new capacity-extended cell, and no further investigation and control is needed to manage the cells inside. The parallel cells will naturally have the same voltage, equal to an embedded cell equalization function based on voltage. In addition, the capacity-extended cell can be modeled as a whole, with larger capacity and smaller internal resistance than the individual cells.

As shown in Fig. 5.16, three cells are parallel-connected to form a new capacity-extended 6.9-Ah cell; then eight new cells are series-connected to form a smart module managed by a module controller. In the module controller, each AD7280 investigates four new cells. Finally, two smart modules are series-connected to form the battery pack controlled by a central module. The capacity-extended cell and the final battery pack are shown in Fig. 5.17.

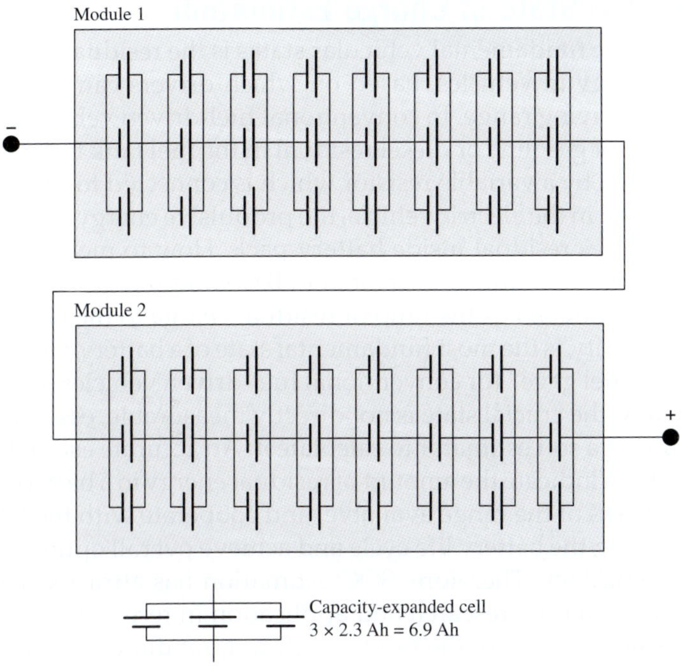

FIGURE 5.16 Grouping architecture of the applied battery pack.

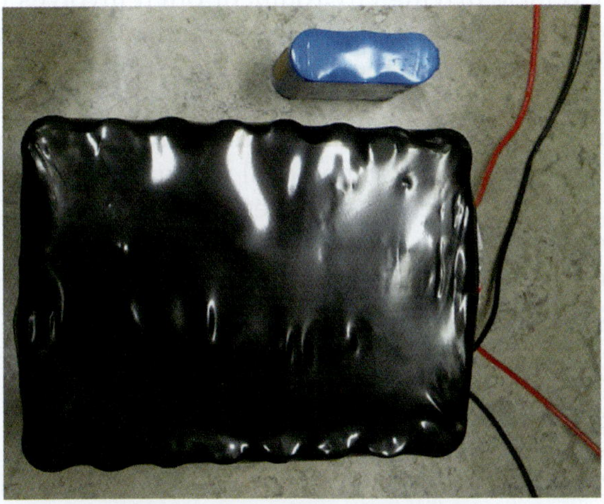

FIGURE 5.17 The capacity-extended cell and battery pack used in this work.

5.2 Robust State of Charge Estimation

One of the fundamental vehicular states is the residual propulsive energy in vehicles, based on which drivers can estimate the left driving range. In conventional fuel-driven vehicles, the amount of gasoline or diesel residual in the fuel tank is directly measured by a variable resistor which is connected to a floater. However, in the electric vehicle, the propulsive energy becomes the charges residual inside battery pack. How to measure the left charges is a difficult problem in BMS design.

SOC, defined as the ratio of residual charge to the nominal full capacity, is the most fundamental state of a battery, working as the "fuel gage" for conventional fuel-driven vehicles. Unfortunately, the crucial state is not directly measurable, essentially requiring a soft estimator to calculate it. An accurate estimation result can indicate the amount of residual energy in a battery, inform users of the range available, and cooperate with the VMS to prolong the battery life cycle and achieve overall optimal energy efficiency. Therefore, SOC estimation has attracted wide coverage in both research and application in recent years, and it has become one of the most significant but difficult issues in BMS design.

5.2.1 Framework Overview

The real vehicle driving environment often involves interference sources that cause signal measurement noise and even zero drift. Meanwhile, the strong time-variant properties of batteries makes it difficult to establish a sufficiently accurate linear model to estimate and predict batteries' dynamic behavior. In other words, the nonideal working conditions make it hard to satisfy the prerequisites of most individual SOC estimators.

Therefore, to guarantee estimation accuracy in the real driving process, it is necessary to improve the antinoise abilities, or robustness, of SOC estimation techniques. In this section, we have proposed a robust estimation framework, as shown in Fig. 5.18, which consists of the following:

1. A nonlinear diffusion filter to reduce current measurement noise, where I_m is the measured current and I_f is the filtering result

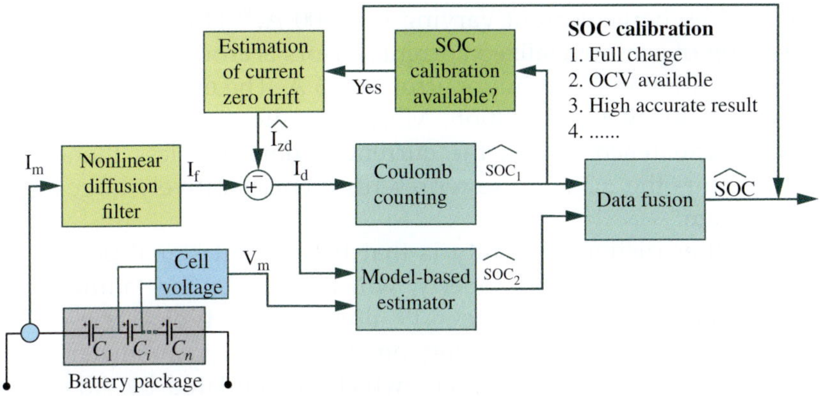

FIGURE 5.18 The proposed robust SOC estimation framework.

2. A current zero-drift estimator to remove the zero drift, where $\widehat{I_{zd}}$ is the estimated zero-drift of current measurement and $I_d = I_f - \widehat{I_{zd}}$ is the denoised result.
3. A coulomb counting estimator to implement open-loop estimation $\widehat{SOC}_1$.
4. A model-based estimator to realize closed-loop estimation, where V_m is the measured terminal voltage and $\widehat{SOC}_2$ is the estimated result.
5. A data fusion component to achieve the final estimation $\widehat{SOC}$ by integrating $\widehat{SOC}_1$ and $\widehat{SOC}_2$.

5.2.2 Current Denoising

A battery is a typical less-information system, where a complex multiparameter electrochemical reaction occurs inside, while only terminal voltage, bus current, and surface temperature can be measured outside. SOC, one of the internal states, has to be estimated according to the limited external variables. Thus, the measurement accuracy of these variables is crucial for SOC estimation.

5.2.2.1 Property Analysis of Current Measurement

A vehicular battery pack usually consists of tens, even hundreds, of series- and parallel-connected cells to generate a large charge/discharge current varying in ±300 A.[36] However the precision of commercialized current sensors is around ±1 percent, resulting in a maximum error of ±3 A. The error is nonignorable for SOC estimation. Moreover, although the peak current can reach 300 A, the current is normally less than 100 A; thus the ±3-A noise represents a comparatively large percentage.

Another distinct difficulty is that the current signal possesses a quasi-random property, which leads to the failure of traditional filters. Figure 5.19 illustrates a typical current profile of a Prius driving on a Urban Dynamometer Driving Schedule (UDDS) cycle, which is simulated by Advisor. Battery current is determined by the demand from the motor and/or generator, and it primarily depends on driving behavior. The way a vehicle is driven is limited by road conditions and may be disrupted by various unexpected events, such as traffic jams, braking to avoid pedestrians, sudden acceleration when overtaking, and so forth. Such erratic driving will inevitably result in quasi-random current

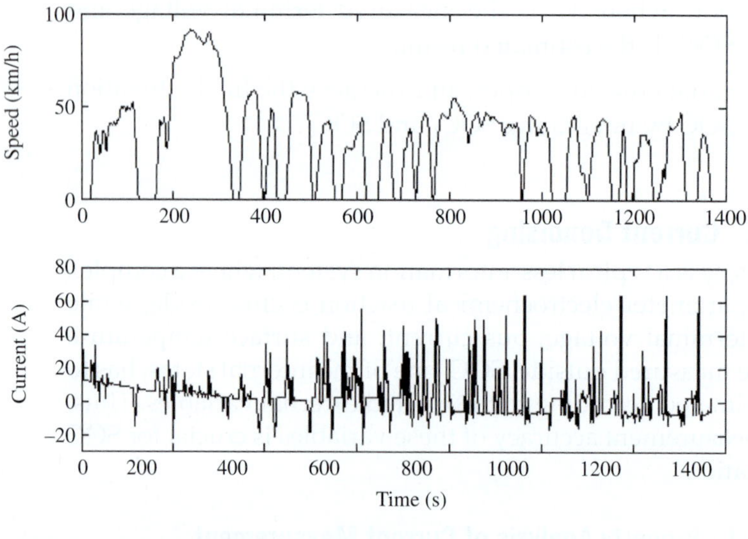

FIGURE 5.19 Current profile of Prius driving on cycle UDDS.

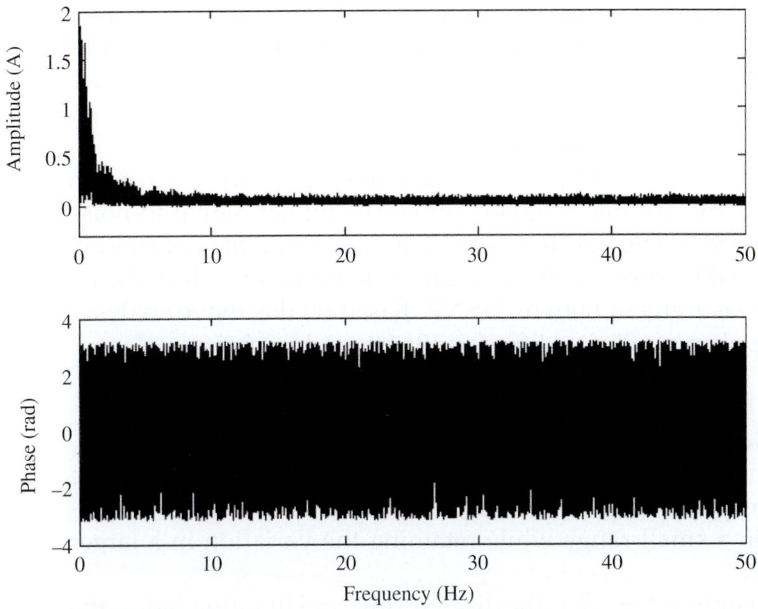

FIGURE 5.20 Frequency analysis of current profile of Prius driving on cycle UDDS.

on the power bus. Analysis of the frequency domain further depicts the property of the battery current, as shown in Fig. 5.20. The current signal, with 100-Hz sampling frequency, expands across the whole frequency domain, and it is hard to determine a cutoff frequency that separates signal from noise.

Noise reduction essentially requires some kind of property difference between signal and noise. As discussed earlier, the similar frequency properties of current signal and noise cause difficulty in applying traditional filters, e.g., low-pass filters, to isolate noise.

The variation in real current signal is caused by driving behavior, which usually leads to a comparatively large degree of change. However, current noise is often produced by the precision of measurement units, electromagnetic interference, vehicle vibrations, and so on, which typically varies in a small range.

According to experiments, we find that the noise in current measurement is continued within a limited region. In addition, the noise is time-independent. Therefore, in this work, we assume the noise has gaussian distribution with a fixed standard deviation and a nonzero mean.

Under this assumption, it is clear that large current variation caused by driving behavior possesses higher signal-to-noise ratio (SNR). The filtering of current measurement is purposely designed to improve the measurement with low SNR and keep the measurement with high SNR. Based on the above analysis, we apply a nonlinear diffusion filter to reduce noise.

5.2.2.2 Nonlinear Diffusion Filter

The nonlinear diffusion filter (NDF) was first proposed in the image processing field to nonlinearly eliminate the oscillation in a small range while retaining the variation in a larger range.[81]

Denoting k and i as the discrete time and iteration index, the iterative equation of the nonlinear diffusion filter is described as

$$\begin{aligned} I(k, i+1) = I(k, i) + \Delta t \times c_l(k, i) \times [I(k-1, i) - I(k, i)] \\ + \Delta t \times c_r(k, i) \times [I(k+1, i) - I(k, i)] \end{aligned} \tag{5.6}$$

where initial iteration $I(k, 1)$ is set as the measured data sequence $I_m(k)$, the last iteration $I(k, N)$ is the filtering result $I_f(k)$, and N is the iteration time.

As an iterative step, a large Δt stands for a long diffusion period in each iteration, leading to smoother results, but with the risk of an unstable iteration process. In contrast, too small a Δt causes a slow diffusion process, requiring more iterations to achieve a satisfactory result.

The $c_l(k, i)$ and $c_r(k, i)$, generally denoted by $c(k, i)$, are the left and right diffusion coefficients, respectively, which control the degree of smoothing for values between the kth datum and its left or right neighbor. The smaller $c(k, i)$ is, the harder it is to smooth. As discussed earlier, a large change in current has a high possibility of being with large SNR, while a small change usually possesses low SNR. Thus, we set $c(k, i)$ to be inversely proportioned to the signal difference, using the conventional equations

$$c_l(k, i) = \frac{1}{1 + \left[\frac{I(k-1,i)-I(k,i)}{\kappa}\right]^2} \tag{5.7}$$

and

$$c_r(k, i) = \frac{1}{1 + \left[\frac{I(k+1,i)-I(k,i)}{\kappa}\right]^2} \tag{5.8}$$

where κ is the gradient modulus threshold that controls the conduction.

5.2.2.3 Adaptive-κ Strategy

Equations (5.7) and (5.8) indicate that the performance of the nonlinear diffusion filter depends on the selection of κ. In general, a noisy signal with low SNR requires a large κ to enhance the smoothing effect. However, the SNR of a noisy signal is an unknown value and has to be estimated indirectly. Based on the characteristics of battery current, an adaptive-κ selection method is proposed using the following steps.

1. Calculate the differential signal $\Delta I_m(k)$ of the original signal $I_m(k)$, where the duration between k and $k+1$ is the sampling period.

$$\Delta I_m(k) = I_m(k+1) - I_m(k) \tag{5.9}$$

2. Establish the set Δ_s in which the absolute values of the differential signal are smaller than 5 A. The reason to remove higher differential data is that they are likely caused by driving behavior, i.e., the real current change.

$$\Delta_s = \{\Delta I_m(k) \mid |\Delta I_m(k)| < 5\} \tag{5.10}$$

3. Calculate the standard deviation SD(Δ_s) of the elements in set Δ_s. To some extent, SD(Δ_s) is an indicator of the SNR of signal. The smaller SD(Δ_s) is, the larger SNR is.

4. Adaptively determine κ according to $SD(\Delta_s)$. In this study, first we fix the iteration time $N = 20 f_s$ (f_s stands for sampling frequency) and the iterative step $\Delta t = 0.6$; then the relationship between κ and $SD(\Delta_s)$ is experimentally determined by the below piecewise function.

$$\kappa = \begin{cases} \min(SD(\Delta_s)/10, 2) & SD(\Delta_s) > 2.2 \\ \max(SD(\Delta_s) \times 1.76 - 3.652, 0) & SD(\Delta_s) < 2.2 \end{cases} \tag{5.11}$$

Remark: As the nonlinear diffusion filter requires the right neighbors (future data) to smooth the current datum, the filter has to delay for a period. In this work, the iteration time $N = 20 f_s$ will result in a 20-s delay, which is short enough to be negligible because SOC changes slowly.

5.2.2.4 Performance Comparison

To demonstrate the efficacy of the nonlinear diffusion filter with adaptive-κ strategy, performance is compared with the traditional low-pass filter and wavelet filter. The low-pass filter is implemented using the Butterworth method with cutoff frequency $f_{cut} = 0.9(f_s/2)$ determined by trial and error. The wavelet filter uses level-dependent thresholds determined by the Birge-Massart strategy.[6]

The current profiles are produced by Advisor. We select the well-developed "Prius_jpn" vehicle model and three typical driving cycles: the highway HWFET, the suburban WVUSUB, and the urban MANHATTAN. The current signal is corrupted by white noise with SNR from 0 to 20 dB.

The performance index of the reduced root-mean-square of errors (RRMSoE) is defined below.

$$\text{RRMSoE}[A] = \text{rms}(I, I_m) - \text{rms}(I, I_f) \tag{5.12}$$

and

$$\text{RRMSoE}[\%] = \frac{\text{rms}(I, I_m) - \text{rms}(I, I_f)}{\text{rms}(I, I_m)} \times 100\% \tag{5.13}$$

where

$$\mathrm{rms}(I_1, I_2) = \sqrt{\frac{1}{L}\sum_{i=1}^{L}[I_1(i) - I_2(i)]^2} \tag{5.14}$$

and I, I_m, and I_f are the real current, measured noisy current, and filtering result, respectively, and L is the length of data.

As an example, Fig. 5.21 illustrates the HWFET case with 15-dB SNR noise. The top figure describes the real current signal and noise signal. It is clear that the real signal has larger variance. Added noise appears as burrs. The errors of the filtering results are given in the bottom figure. Without doubt, the low-pass filter gives poor performance, even worse than the noisy signal, because it simultaneously removes the real signal in the high-frequency zone. Basically, any frequency-based filter will have difficulty handling this problem. The wavelet filter, focusing on both scale and time, performs better than the low-pass filter but still fails to keep the large variation in the

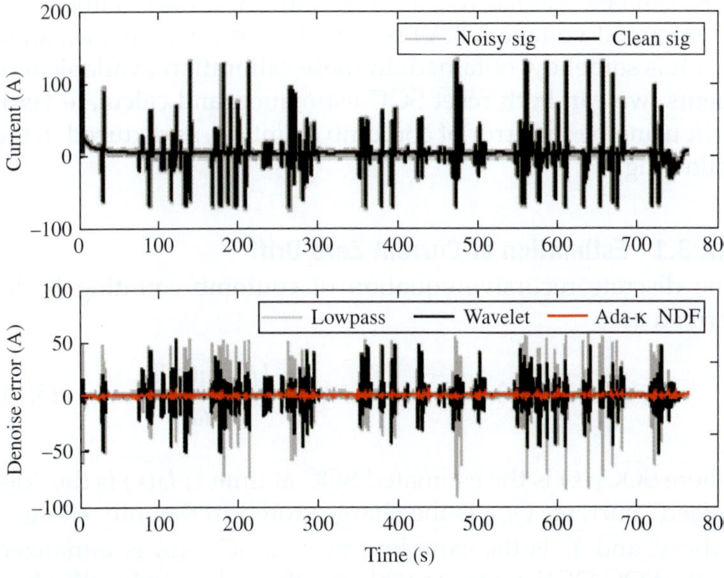

FIGURE 5.21 Filtering results of HWFET current profile corrupted by noise with 15-dB SNR.

real signal. The nonlinear diffusion filter successfully removes the noise appearing as burrs while keeping the real signal with large variation in morphology.

The total results are given in Table 5.1, where each value is the average of 100 independent random tests. It is clear that the adaptive-κ strategy efficiently estimates the equivalent SNR in each case and hence calculates a suitable κ to achieve satisfactory denoising.

5.2.3 Current Zero-Drift Reduction

Although the nonlinear diffusion filter has the ability to remove the oscillatory noise, it cannot handle the zero-drift problem which causes the baseline shift. Zero drift, actually the nonzero mean of noise, is the source of divergent results in coulomb counting, which negatively affects the performance of other estimators.

In this section, we propose a self-learning strategy to estimate the zero drift of current measurement. To establish the self-learning system, it is necessary to know the estimation error, equally the real SOC. In practice, the true value of the SOC can be obtained when (1) the battery is fully charged, (2) open-circuit voltage (OCV) is available, or (3) a more accurate result is somehow obtained. In these calibration-available moments, we can both reset SOC estimation and calculate zero-drift using by the error of coulomb counting, as deduced in the following.

5.2.3.1 Estimation of Current Zero Drift

The discrete recursive equation of coulomb counting is described as

$$\widehat{\mathrm{SOC}}_1(k) = \widehat{\mathrm{SOC}}_1(k-1) + \frac{I_d(k)\eta(k)T_s}{Q_{\mathrm{full}}} \tag{5.15}$$

where $\widehat{\mathrm{SOC}}_1(k)$ is the estimated SOC at time k; $I_d(k)$ is the "denoised" current; Q_{full} is the charge stored in the fully charged battery, and T_s is the sampling period. $\widehat{\mathrm{SOC}}_1(0)$ is initialized by the SOC-OCV mapping table or other advanced methods at each start time, or it is reset to the true value at each calibration-available time.

SNR	Highway (HWFET)						Suburban (WVUSUB)						Urban (MANHATTAN)					
	Lowpass		Wavelet		Ada-κ NDF		Lowpass		Wavelet		Ada-κ NDF		Lowpass		Wavelet		Ada-κ NDF	
(dB)	A	%	A	%	A	%	A	%	A	%	A	%	A	%	A	%	A	%
0	−2.05	−21.98	2.98	31.95	**5.53**	**59.41**	0.00	−0.04	4.82	52.09	**6.68**	**72.15**	−0.14	−1.22	5.57	49.58	**6.76**	**60.23**
2	−2.62	−35.47	1.57	21.26	**4.11**	**55.53**	−0.24	−3.29	3.85	52.35	**5.33**	**72.53**	−0.43	−4.83	4.33	48.50	**5.77**	**64.69**
4	−3.14	−53.41	0.41	7.02	**2.82**	**47.93**	−0.44	−7.58	3.07	52.52	**4.04**	**69.25**	−0.70	−9.85	3.27	46.15	**4.32**	**60.94**
6	−3.77	−80.85	−0.58	−12.34	**1.86**	**39.84**	−0.67	−14.52	2.42	52.15	**2.99**	**64.44**	−0.99	−17.66	2.35	41.75	**3.10**	**55.14**
8	−4.24	−114.45	−1.42	−38.27	**1.08**	**29.03**	−0.91	−24.73	1.88	51.01	**2.18**	**59.30**	−1.33	−29.67	1.53	34.28	**2.11**	**47.13**
10	−4.73	−160.50	−2.11	−71.68	**0.71**	**24.00**	−1.16	−39.61	1.44	49.30	**1.58**	**54.08**	−1.63	−45.89	0.86	24.24	**1.44**	**40.53**
12	−5.16	−220.51	−2.67	−114.30	**0.70**	**29.84**	−1.40	−60.15	1.06	45.47	**1.29**	**55.32**	−1.96	−69.49	0.28	10.00	**1.06**	**37.72**
14	−5.51	−296.75	−3.12	−168.11	**0.94**	**50.74**	−1.63	−88.11	0.72	39.25	**1.20**	**64.90**	−2.24	−100.18	−0.21	−9.19	**0.93**	**41.54**
16	−5.81	−393.55	−3.49	−236.44	**0.80**	**54.51**	−1.84	−125.45	0.43	29.53	**0.93**	**63.17**	−2.52	−141.40	−0.63	−35.24	**1.00**	**56.34**
18	−6.07	−517.99	−3.78	−322.74	**0.63**	**53.77**	−2.03	−174.06	0.18	15.67	**0.73**	**62.35**	−2.76	−195.17	−0.96	−67.62	**0.76**	**53.99**
20	−6.28	−674.39	−4.02	−431.19	**0.49**	**52.57**	−2.20	−237.73	−0.02	−2.45	**0.57**	**62.03**	−2.96	−263.57	−1.23	−109.44	**0.59**	**52.19**

TABLE 5.1 Reduced rms of Denoising Errors Based on Different Filters

The $\eta(k)$ is the coulombic efficiency or ampere-hour efficiency. Strictly speaking, it is a time-variant parameter, depending on temperature, SOC, and other relative states. The basic way of determining coulombic efficiency is to establish its value table according to the manufacturers' datasheets or testing data.[48] The mass utilization of coulomb counting in practical applications has demonstrated its feasibility and validity. In addition, adaptive learning and online estimation strategies can strengthen the accuracy of coulombic efficiency estimation.[51, 69, 113] Moreover, since the accuracy of the coulomb counting primarily depends on measurement of the battery current and estimation of the initial SOC,[69] the error of coulombic efficiency will not be covered in this section.

To analyze the zero drift of current measurement, we denote the true value of current as $I(k)$, the residual noise affecting the coulomb counting estimator as $I_n(k)$, and the true value of SOC as SOC(k). The "denoised" current accumulated in the coulomb counting method is expressed by

$$I_d(k) = I(k) + I_n(k) \tag{5.16}$$

and

$$\begin{aligned}
\widehat{\mathrm{SOC}}_1(k) &= \widehat{\mathrm{SOC}}_1(k-1) + \frac{I_d(k)\eta(k)T_s}{Q_{\mathrm{full}}} \\
&= \widehat{\mathrm{SOC}}_1(0) + \frac{\sum_{i=0}^{k} I_d(k)\eta(i)T_s}{Q_{\mathrm{full}}} \\
&= \widehat{\mathrm{SOC}}_1(0) + \frac{T_s\left[\sum_{i=0}^{k} I(i)\eta(i) + \sum_{i=0}^{k} I_n(i)\eta(i)\right]}{Q_{\mathrm{full}}} \\
&= \left[\widehat{\mathrm{SOC}}_1(0) + \frac{T_s\sum_{i=0}^{k} I(i)\eta(i)}{Q_{\mathrm{full}}}\right] + \frac{T_s\sum_{i=0}^{k} I_n(i)\eta(i)}{Q_{\mathrm{full}}}
\end{aligned} \tag{5.17}$$

Since $\widehat{\mathrm{SOC}}_1(0)$ is initialized at each start time by the SOC-OCV mapping table or advanced algorithms, or is reset to the "true" value at each calibration-available moment, its error is small enough to be ignored within the tolerance range. Therefore we have

$$\widehat{\mathrm{SOC}}_1(0) = \mathrm{SOC}(0) \tag{5.18}$$

$$\widehat{\mathrm{SOC}}_1(k) = \mathrm{SOC}(k) + \frac{T_s \sum_{i=0}^{k} I_n(i)\eta(i)}{Q_{\mathrm{full}}} \tag{5.19}$$

$$E\left[\widehat{\mathrm{SOC}}_1(k) - \mathrm{SOC}(k)\right] = E\left[\frac{T_s \sum_{i=0}^{k} I_n(i)\eta(i)}{Q_{\mathrm{full}}}\right]$$

$$= \frac{T_s \sum_{i=0}^{k} \eta(i)}{Q_{\mathrm{full}}} E\left[I_n(i)\right] \tag{5.20}$$

At each calibration-available time k^*, the true value $\mathrm{SOC}(k^*)$ is available; thus

$$\widehat{\mathrm{SOC}}_1(k^*) - \mathrm{SOC}(k^*) = \frac{T_s \sum_{i=0}^{k^*} \eta(i)}{Q_{\mathrm{full}}} E\left[I_n(i)\right] \tag{5.21}$$

The zero drift of current measurement $\widehat{I_{zd}}$ is the mean of noise:

$$\widehat{I_{zd}} = E\left[I_n(i)\right] = \frac{Q_{\mathrm{full}}\left[\widehat{\mathrm{SOC}}_1(k^*) - SOC(k^*)\right]}{T_s \sum_{i=0}^{k^*} \eta(i)} \tag{5.22}$$

At each calibration-available moment, $\widehat{I_{zd}}$ is updated by Eq. (5.22), which enables online tracking of zero drift.

5.2.3.2 Performance Comparison

To demonstrate availability and efficacy, simulations based on Advisor have been conducted to track SOC when the

Prius drives on the cycles HWFET, WVUSUB, MANHATTAN, NEDC, US06, and NYCC successively. After completion of each cycle, a half-hour stop allows the battery OCV to be determined. When the vehicle starts a new driving cycle, SOC is reset to the real value and the algorithm reestimates zero drift. The SNR of current measurement is set to 0 dB. The zero drift I_{zd} is set to 0.5 A in the first through third cycles and 1 A in the fourth through sixth cycles. BMS first denoises the current by ada-κ NDF and then removes the zero drift by subtracting the estimation value $\widehat{I_{zd}}$.

Figure 5.22 shows the SOC tracking results, comparing performance either using or not using the zero-drift self-learning strategy. Table 5.2 summarizes the numerical indexes of each cycle, including the average (ave) of absolute SOC estimation errors and the standard deviation (SD) of errors in each cycle. Since we have no prior knowledge about zero drift, its estimation value is set to 0 in the first driving cycle, HWFET; it therefore achieves the same SOC tracking performance as with the conventional method. In the following two cycles, OCV is available at each start time to calibrate the SOC and to calculate the estimation error of the previous cycle, which allows $\widehat{I_{zd}}$ to

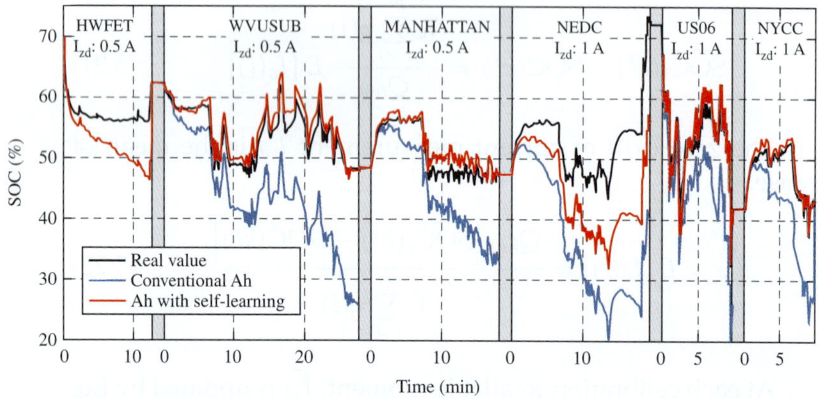

FIGURE 5.22 SOC estimation results based on conventional coulomb counting and the proposed coulomb counting with self-learning strategy.

The gray bars between cycles represent half-hour stops which allow OCV available. Simulation conditions: white noise with 0-dB SNR, zero drift of 0.5 A for the first through third cycles and 1 A for the fourth through sixth cycles.

SOC Estimation Methods	HWFET		WVUSUB		MANHATTAN		NEDC		US06		NYCC	
	ave	SD	ave	SD	ave	SD	ave	SD	ave	SD	ave	SD
Conventional Ah	4.71	2.71	9.87	6.22	6.20	4.04	16.21	9.36	7.74	4.36	7.16	4.86
Ah with self-learning	4.71	2.71	1.34	0.96	1.23	0.95	8.24	4.77	1.42	1.65	0.96	0.50

TABLE 5.2 Numerical Results of SOC Tracking Using or Not Using Self-Learning Strategy (Unit [%])

update. The estimation values of zero drift are 0.4895 and 0.4952 A, respectively, which are very close to the real zero drift of 0.5 A. Thus, the estimation error of the proposed method is clearly smaller than that of the conventional algorithm. In the fourth cycle, the real zero drift is changed to 1 A, with the aim of testing the self-adaptive ability of the learning strategy. As the learning strategy determines zero drift by using the error in the previous cycle, the one-cycle delay means that $\widehat{I_{zd}} = 0.4987$ cannot fully compensate the real 1-A drift, resulting in nonconvergent SOC estimation. In the following cycles, the large error updates $\widehat{I_{zd}}$ to be 0.9863 and 0.9967 in the fifth and sixth cycles, respectively. The self-adaptive ability leads to satisfactory performance in the last two cycles.

Although the proposed framework significantly improves on the performance of the conventional coulomb counting method, it is inherently an open-loop estimator that does not take the measurable voltage into consideration. It only self-corrects estimation results at calibration moments but cannot revise errors during the driving process.

5.2.4 Simulation: RC Model and H_∞ Filter

A battery is a typical time-variant system, closely related to SOC and other battery states. Online identification is the popular strategy for solving the time-variant problem, but at the cost of high time consumption and inaccurate identification results due to noise. An alternative method is to use a robust estimation technique, which constructs a suboptimal filter with the ability to minimize the maximum estimation error caused by the noise and uncertainties of the system model.

In pace with the development of H_∞ control theory, researchers have shown great interest in the H_∞ filter.[32, 33] A good introduction and review can be found in Ref. 100. In contrast to Kalman filter, the H_∞ filter is proposed to handle estimation problems under conditions of uncertain model structure, model parameters, and system noise. It has two main features: (1) it does not require any assumptions about disturbances and model uncertainties; (2) it minimizes the estimation error in the worst situation. Therefore, it is more applicable than the Kalman filter in practical applications.

5.2.4.1 H_∞ Filter Algorithm

Denoting x as the system state vector, y the output vector, u the input vector, w the process noise, and v the measurement noise, system state space equations are expressed as

$$\begin{aligned} x_{k+1} &= A_k x_k + B_k u_k + \Gamma_k w_k \\ y_k &= C_k x_k + D_k u_k + v_k \end{aligned} \tag{5.23}$$

The suboptimal H_∞ filtering problem is formalized as follows: given the estimation error bounder $\gamma > 0$, find an estimation of $\hat{z}_k = F_f(u_0, \ldots, u_k, y_0, \ldots, y_k)$ such that

$$\inf_{F_f} \sup_{x_0, w \in H_2, v \in H_2} \frac{\|z_k - \hat{z}_k\|_2^2}{\|x_0 - \hat{x}_0\|_{P_0^{-1}}^2 + \|w_k\|_2^2 + \|v_k\|_2^2} < \gamma^2 \tag{5.24}$$

where $z_k = L_k x_k$ is the estimation goal, i.e., the linear combination of system states, L_k is the user-defined state weight matrix, and $\|\chi_k\|_2^2 = \sum_{i=0}^{k} \chi_i^* \chi_i$.

The solution of the suboptimal H_∞ filtering problem can be calculated by the following recursion formulas:

$$\hat{x}_{k+1,k} = A_k \hat{x}_k + B_k u_k \tag{5.25}$$

$$R_k = \begin{bmatrix} I & 0 \\ 0 & -\gamma^2 I \end{bmatrix} + \begin{bmatrix} C_k \\ L_k \end{bmatrix} P_k \begin{bmatrix} C_k^T & L_k^T \end{bmatrix} \tag{5.26}$$

$$P_{k+1} = A_k P_k A_k^T + \Gamma_k \Gamma_k^T - A_k P_k [C_k^T \, L_k^T] R_k^{-1} \begin{bmatrix} C_k \\ L_k \end{bmatrix} P_k A_k^T \tag{5.27}$$

$$K_{k+1} = P_{k+1} C_{k+1}^T \left[I + C_{k+1} P_{k+1} C_{k+1}^T \right]^{-1} \tag{5.28}$$

$$\hat{x}_{k+1} = \hat{x}_{k+1,k} + K_{k+1}(y_{k+1} - C_{k+1}\hat{x}_{k+1,k} - D_{k+1} u_{k+1}) \tag{5.29}$$

$$\hat{z}_{k+1} = L_{k+1} \hat{x}_{k+1} \tag{5.30}$$

under the conditions that $[A_k\ \Gamma_k]$ has full rank and

$$P_k^{-1} + C_k^T C_k - \gamma^{-2} L_k^T L_k > 0 \qquad \text{for all } k \tag{5.31}$$

From the above formulas, it is clear that if $L(k) = I$ and $\gamma \to \infty$, the H_∞ filter degrades to be the Kalman filter. Thus, the Kalman filter is a special situation of the H_∞ filter with infinite H_∞ norm, and hence it is the least robust.

5.2.4.2 Battery *RC* Model

As shown in Fig. 5.23, we apply the battery *RC* equivalent circuit model in this chapter.[42] The *RC* model consists of a bulk capacitor C_b and a surface capacitor C_s, which simulate energy storage and the dynamic property of the battery, respectively. Output resistance R_o, surface resistance R_s, and bulk resistance R_b are used to model the internal resistance of the battery.

By selecting the state vector as the voltages of bulk and surface capacitors $x_k = [V_{bk}, V_{sk}]'$, system input as bus current $u_k = I_k$, output as terminal voltage $y_k = V_{ok}$, and sampling time as T_s, the discrete state space Eqs. (5.23) are concrete in the following matrixes.

$$A = \begin{bmatrix} 1 - \dfrac{T_s}{C_b(R_b + R_s)} & \dfrac{T_s}{C_b(R_b + R_s)} \\ \dfrac{T_s}{C_s(R_b + R_s)} & 1 - \dfrac{T_s}{C_s(R_b + R_s)} \end{bmatrix} \tag{5.32}$$

$$B = \begin{bmatrix} \dfrac{-R_s T_s}{C_b(R_b + R_s)} \\ \dfrac{-R_b T_s}{C_s(R_b + R_s)} \end{bmatrix} \tag{5.33}$$

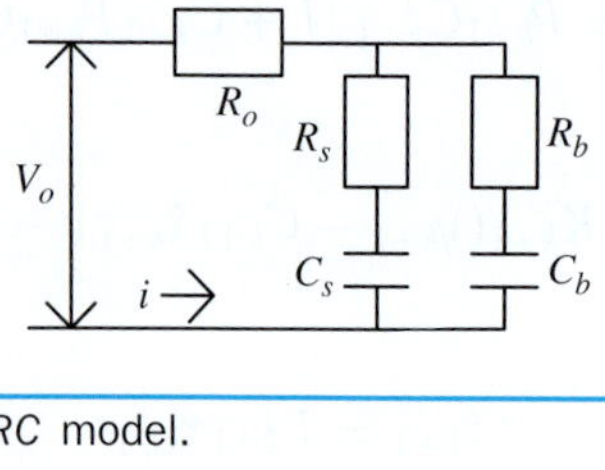

FIGURE 5.23 Battery *RC* model.

$$C = \left[\frac{R_s}{R_b + R_s} \quad \frac{R_b}{R_b + R_s} \right] \tag{5.34}$$

$$D = -R_o - \frac{R_b R_s}{R_b + R_s} \tag{5.35}$$

The SOC for the *RC* model was estimated by using the voltages of the two capacitors. Since C_b represents the bulk energy in the battery, it contributes the majority of SOC, as expressed in the equations below.

$$\widehat{\mathrm{SOC}}_2(k) = \frac{1}{21} \left[20\widehat{\mathrm{SOC}}_{C_b}(k) + \widehat{\mathrm{SOC}}_{C_s}(k) \right] \tag{5.36}$$

where

$$\widehat{\mathrm{SOC}}_{C_b}(k) = F_{\mathrm{OCV-SOC}}(V_{bk}) = F_{\mathrm{OCV-SOC}}(x_k(1)) \tag{5.37}$$

$$\widehat{\mathrm{SOC}}_{C_s}(k) = F_{\mathrm{OCV-SOC}}(V_{sk}) = F_{\mathrm{OCV-SOC}}(x_k(2)) \tag{5.38}$$

and $F_{\mathrm{OCV-SOC}}(\cdot)$ is the function mapping OCV to SOC. It usually is predetermined the manufacturer's datasheet or experimental testing data.

Model parameters will essentially change during the running process. A type of 6.5-Ah Prismatic Panasonic NiMH battery has been tested at the NREL Battery Thermal Management Lab, and the corresponding model parameters are provided by Advisor.[61] Figure 5.24 shows the change in resistance of the resistors versus temperature and SOC. Taking the output resistor as an example, it is clear that the maximum resistance is 0.0216 Ω, more than three times the minimum value 0.0071 Ω. The significant variation in the model parameters is worth paying special attention to in the design of the SOC estimator. Other resistors and capacitors have similar properties.

Since the model error is hard to determine beforehand, we leave this difficulty to the H_∞ filter and estimate Γ only according to the current noise. The measured current is $u = I_c = I + I_n$, where I is the clean signal and I_n is the current noise. Therefore, $Bu = BI + BI_n = BI + Bw$, that is, $\Gamma = B$.

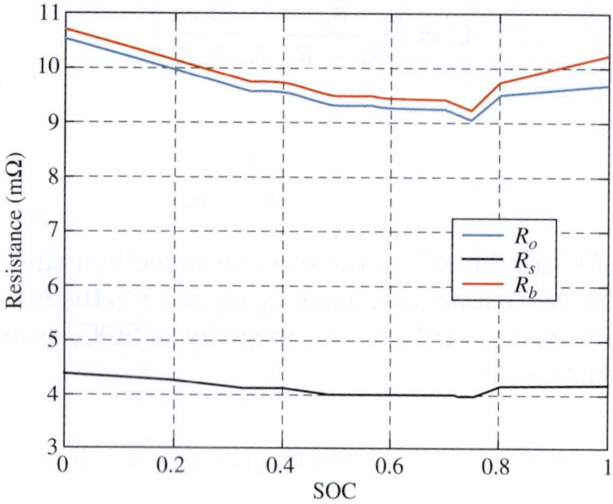

FIGURE 5.24 Resistance variance of resistors in *RC* model.

5.2.4.3 Performance Comparison

To demonstrate the performance of the H_∞ filter, simulation experiments are conducted to compare the SOC estimation errors among the model output, the results of the Kalman filter, and the results of the H_∞ filter.

As demonstrated earlier, the adaptive-κ nonlinear diffusion filter has the ability to remove the current noise, and a self-learning strategy can compensate for zero drift. Therefore, we fix the SNR of current noise at 10 dB and zero drift at 0.5 A, which simulates the residual noise after the two noise reduction steps. The experiments aim to verify the filters' abilities to handle the modeling error; thus we apply a fixed model with the parameters set to their maximum, average, and minimum values of each cycle. The inaccurate model parameters include the resistances of R_b, R_s, R_o and capacitance of C_c. Since C_b is determined by the nominal capacity of the battery, rather than identification, no modeling error is added to it.

Battery model parameters are from the Panasonic Prismatic 6.5-Ah battery. The settings for the Kalman filter are experimentally optimized as initial states estimation error $P = 0.01I$, process noise variance matrix $Q = 0.001^2 I$, and measurement noise variance matrix $R = 0.01^2$. The settings for the H_∞ filter

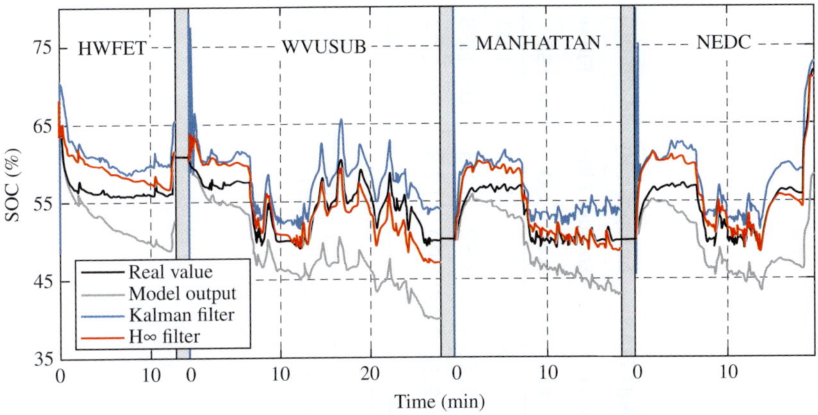

FIGURE 5.25 SOC estimation results based on inaccurate battery model using the fixed average values of real time-variant parameters.

The gray bars between cycles represent half-hour stops which allow the OCV available. Simulation conditions: white noise with 10-dB SNR and 0.5-A zero drift.

are $L = [1, 1]$ and $\gamma = 50$. If γ fails to satisfy the condition in Eq. (5.31), it will increase 10 step by step until it meets the requirement.

Figure 5.25 shows the SOC estimation results of the Prius, driving on successive HWFET, WVUSUB, MANHATTAN, and NEDC cycles. Because of the nonzero mean of current noise and model error, the accumulated total errors are reflected in the model output, resulting in a nonconvergent estimation result. After a short time of oscillation, the Kalman filter gradually converges to a stable estimation, but with stable errors. Zero-drift and model errors destroy its ability to act as an optimal filter. The H_∞ filter leads to a faster convergence process than does the Kalman filter and achieves fewer stable errors.

Table 5.3 summarizes the numerical results. The H_∞ filter clearly has the ability to estimate SOC of the time-variant battery based on a set of fixed parameters. It outperforms the Kalman filter no matter which values the model parameters are fixed to. The robustness of the H_∞ filter is well demonstrated.

Fix Model Parameters to	Highway (HWFET)				Suburban (WVUSUB)				Urban (MANHATTAN)				European Hybrid (NEDC)			
	Kalman Filter		H_∞ Filter		Kalman Filter		H_∞ Filter		Kalman Filter		H_∞ Filter		Kalman Filter		H_∞ Filter	
	ave	SD	ave	SD	ave	SD	ave	SD	ave	SD	ave	SD	ave	SD	ave	SD
Max. of parameters	4.38	1.41	2.92	2.03	3.99	1.44	1.71	1.59	4.00	1.98	1.35	2.32	3.57	1.22	1.60	1.61
Ave. of parameters	4.34	2.22	2.47	1.21	3.56	1.48	1.94	2.20	3.71	1.88	1.55	1.42	3.82	1.85	1.93	1.83
Min. of parameters	3.77	3.73	1.50	1.15	3.20	1.20	1.78	2.01	3.92	2.24	2.15	2.19	3.90	1.58	2.22	2.34

TABLE 5.3 Performance Comparison of SOC Estimation Based on Battery Model Using a Set of Fixed Parameters (Unit [%])

The inaccurate model parameters include resistances of R_e, R_t, and R_o and capacitance of C_c, and they are fixed to their maximum, average, and minimum values in each driving cycle, respectively. White noises with 10-dB SNR and 0.5-A zero drift are also added to the real current profile.

5.2.5 Experiment and Application: Modified ESC Model and Extended Kalman Filter

5.2.5.1 Modified ESC Model

In experiments and applications, we select and revise the nonlinear battery model ESC proposed in Ref. 84 for two reasons: (1) it is more accurate to model the $LiFePO_4$ battery with a clear physical background; (2) we wish to demonstrate the proposed framework in various situations.

The ESC model is expressed as

$$\begin{bmatrix} f_{k+1} \\ h_{k+1} \\ \mathrm{SOC}_{k+1} \end{bmatrix} = \begin{bmatrix} \mathrm{diag}(\alpha) & 0 & 0 \\ 0 & F(I_k) & 0 \\ 0 & 0 & 1 \end{bmatrix} \begin{bmatrix} f_{k+1} \\ h_{k+1} \\ \mathrm{SOC}_{k+1} \end{bmatrix} + \begin{bmatrix} 1 & 0 \\ 0 & 1 - F(I_k) \\ -\dfrac{\eta_i T_s}{C_n} & 0 \end{bmatrix} \begin{bmatrix} I_k \\ M(\mathrm{SOC}, \dot{\mathrm{SOC}}) \end{bmatrix} \tag{5.39}$$

$$V_k = \mathrm{OCV}(\mathrm{SOC}_k) - RI_k + h_k + Gf_k \tag{5.40}$$

The model first estimates SOC based on the coulombic counting method and then establishes the relationship between SOC and the simultaneous terminal voltage. Since terminal voltage is measurable during the vehicle running process, it is applied as the system output to establish the closed-loop estimation.

The mapping from SOC to terminal voltage consists of four parts: (1) OCV; (2) voltage decrease or increase due to the internal resistance during charge and discharge; (3) hysteresis, which represents the hysteresis phenomenon; and (4) the current filter, which expresses the dynamic behavior of the battery.

In this chapter, we have established the ESC model of the batteries used in experiments, with modification to the implementation of the current filter.

SOC versus OCV The mapping from SOC to OCV $V_{\mathrm{OCV}}(k) = \mathrm{OCV}(\mathrm{SOC}(k))$ is established using a curve-fitting method based on the real data. As shown in Fig. 5.26, we measure the OCV at each concerned SOC point and then apply a nonlinear model to fit the curve.

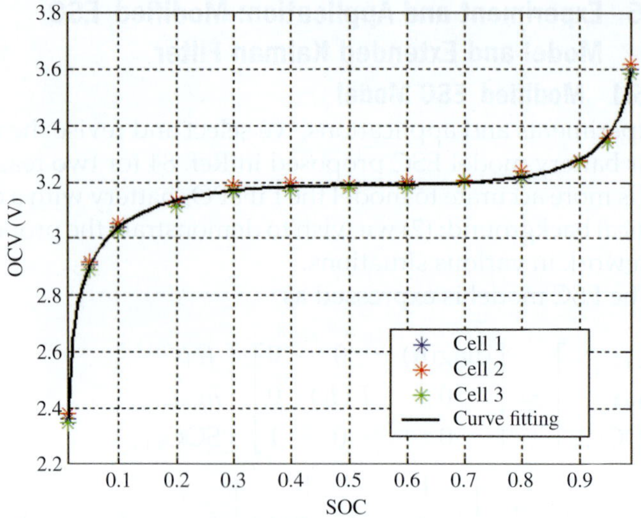

FIGURE 5.26 SOC versus OCV samples and curve-fitting results.

As suggested in Ref. 84, the OCV function is designed as the equation below, to enable its derived function to be calculated easily.

$$\mathrm{OCV}(\mathrm{SOC}(k)) = a_0 - \frac{a_1}{\mathrm{SOC}(k)} - a_2\mathrm{SOC}(k) + a_3 \ln(\mathrm{SOC}(k)) + a_4 \ln(1 - \mathrm{SOC}(k)) \tag{5.41}$$

where $a = [3.545, 0.00338, 0.6605, 0.1828, -0.1539]$, identified by curve fitting.

It is worth noting that the above equation cannot solve the situations SOC = 1 and SOC = 0; therefore we use SOC = 0.01 and SOC = 0.99 to present the empty and full states. In the following, we use the same equation to model internal resistance and hysteresis.

Ohmic Drop/Rise When the battery is operating, discharge or charge current will cause the terminal voltage to rise or fall due to its internal resistance R, as expressed by $RI(k)$.

In a strict sense, the internal discharging and charging resistances of the battery are variables that change along with SOC,

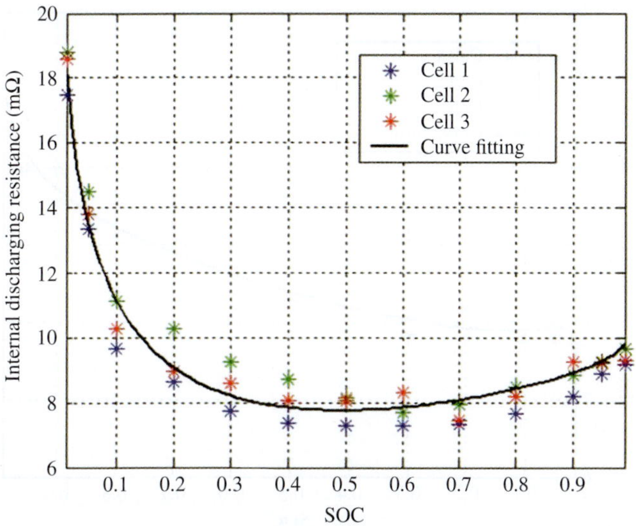

FIGURE 5.27 SOC versus internal discharging resistances of cells.

as shown in Figs. 5.27 and 5.28. Although in Ref. 84 internal resistance is modeled as a constant, we find it is not accurate enough for the application, where the highest internal resistance is more than twice the lowest. Thus, we use the same equation to fit the curve for internal resistance.

For discharging resistance, we have

$$R_{\text{dis}}(\text{SOC}(k)) = a_0 - \frac{a_1}{\text{SOC}(k)} - a_2\text{SOC}(k) + a_3 \ln(\text{SOC}(k)) + a_4 \ln(1 - \text{SOC}(k)) \tag{5.42}$$

where $a = [0.5291, 0.0247, -8.164, -4.341, -0.2616]$, identified by curve fitting.

For charging resistance, we also have

$$R_{\text{chg}}(\text{SOC}(k)) = a_0 - \frac{a_1}{\text{SOC}(k)} - a_2\text{SOC}(k) + a_3 \ln(\text{SOC}(k)) + a_4 \ln(1 - \text{SOC}(k)) \tag{5.43}$$

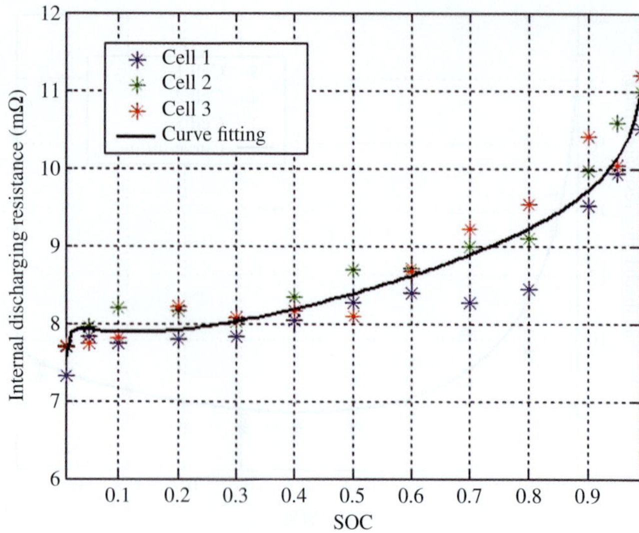

FIGURE 5.28 SOC versus internal charging resistances of cells.

where $a = [6.837, 0.01123, -1.901, -0.3976, -0.4924]$, also identified by curve fitting.

Hysteresis Ideally, after operation of the battery, the terminal voltage will gradually converge to the OCV. However, in practice, the convergence value after a discharge process is always smaller than the OCV, while that after a charge is always bigger. This phenomenon is called *hysteresis*.

To model the hysteresis, the cells are first discharged using a $C/20$ current from full to empty and then charged to full using the same current rate, at 25°C. The two discharge and charge curves, indexed by SOC, are not coincidental. The voltage difference consists of the ohmic voltage and hysteresis. Since the current is very small, the ohmic voltage is only about 2.8 mV. Therefore, the remaining voltage difference is caused by hysteresis. Figure 5.29 shows the half voltage difference caused by the hysteresis effect, with $RI(k)$ subtracted. In this work, we use the same equation to fit the hysteresis curve .

$$M(\mathrm{SOC}(k)) = a_0 - \frac{a_1}{\mathrm{SOC}(k)} - a_2\mathrm{SOC}(k) + a_3 \ln(\mathrm{SOC}(k)) + a_4 \ln(1 - \mathrm{SOC}(k)) \tag{5.44}$$

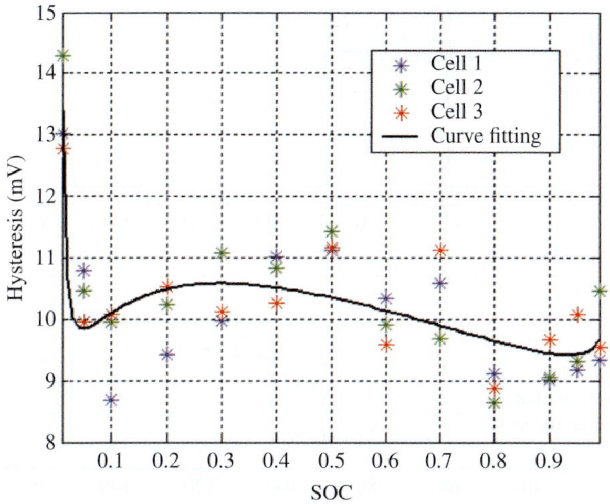

FIGURE 5.29 SOC versus hysteresis of cells.

where $a = [14.39, -0.08124, 6.033, 1.963, -0.2637]$, identified by curve fitting.

In application, $h(k) = s_H(k)M$, where $s_H(k)$ represents the sign of the current and is expressed as

$$s_H(k) = \begin{cases} 1 & I(k) > \varepsilon \\ -1 & I(k) < -\varepsilon \\ s_H(k-1) & |I(k)| < \varepsilon \end{cases} \tag{5.45}$$

Dynamics of Current The above modeling components do not take into consideration the dynamic response of the battery. In fact, the battery takes some time for its terminal voltage to reach a stable value if given a step signal.

As shown in Fig. 5.30, the above model without the consideration of the time constant is unable to simulate the dynamics of the battery. The original method used to simulate the system proposed in ESC applies a state space form $V_f(k) = Gf(k) = GA_f f(k-1) + GB_f I(k-1)$, where $A_f = \text{diag}(\alpha)$ is the state-transition matrix, B_f is the arbitrary input matrix,

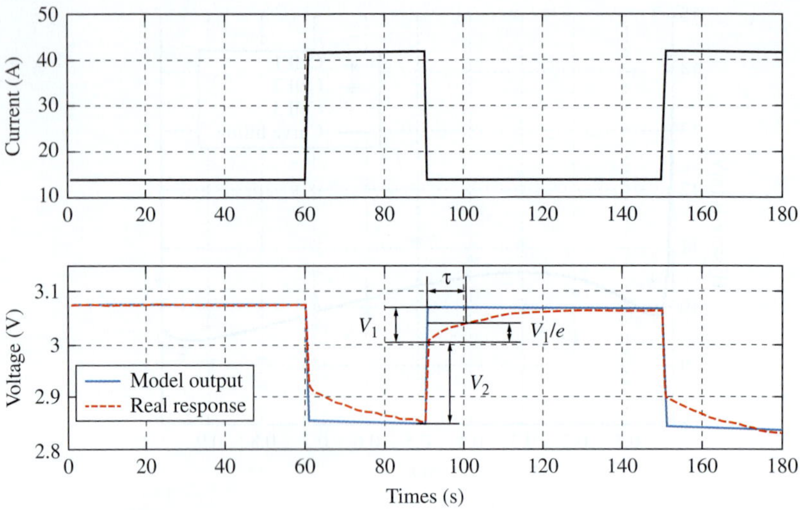

FIGURE 5.30 Battery dynamic response to step discharge current and model output without dynamic consideration.

and G is the output which makes the $Gf(k) \to 0$, during the constant-current period.

However, the design of A_f, B_f, and hence G is not based on any obvious physical meaning. The time constant and gain of the current filter are also not explicitly expressed in the matrix parameters. Since the transitional period is very similar with a exponential decay function, we propose a new design method to realize the modeling of battery dynamics.

Given a step function $u(t)$, the desired response of the battery is an exponential decay $y(t) = e^{-t/\tau}$; therefore, the transfer function can be implemented using the Laplace transform

$$G(s) = \frac{I_f(s)}{I(s)} = \frac{Ks}{s + \tau} \tag{5.46}$$

where K is the gain and τ is the exponential time constant representing the time needed for the initial value to be reduced by a factor of e.

As shown in Fig. 5.30, K is determined by $K = V_1/(V_1 + V_2)$ and τ is measured directly from the duration of time when the initial value V_1 decays to V_1/e. In this chapter, $K = 0.287$ and $\tau = 12.5$ s.

In application, the continuous transfer function is discretized at the sampling period T_s. In this work, discretization is implemented based on a triangle approximation. The discrete-time form is expressed by

$$G(z) = \frac{I_f(z)}{I(z)} = \frac{n_1 z + n_2}{d_1 z + d_2} \tag{5.47}$$

where $n_1 = 0.2758$, $n_2 = -0.2758$, $d_1 = 1$, and $d_2 = -0.9231$ in this work when $T_s = 1$ s.

Finally the whole battery model is given by

$$\begin{bmatrix} I_f(k) \\ \mathrm{SOC}(k) \end{bmatrix} = \begin{bmatrix} -{d_2}/{d_1} & 0 \\ 0 & 1 \end{bmatrix} \begin{bmatrix} I_f(k-1) \\ \mathrm{SOC}(k-1) \end{bmatrix} + \begin{bmatrix} {n_2}/{d_1} & {n_1}/{d_1} \\ 0 & -{\eta T_s}/{C_n} \end{bmatrix} \begin{bmatrix} I(k-1) \\ I(k) \end{bmatrix} \tag{5.48}$$

$$\begin{aligned} V(k) = {} & \mathrm{OCV}(\mathrm{SOC}(k)) - s_{\mathrm{dis}}(k) R_{\mathrm{dis}}(\mathrm{SOC}(k))(I(k) - I_f(k)) \\ & - s_{\mathrm{chg}}(k) R_{\mathrm{chg}}(\mathrm{SOC}(k))(I(k) - I_f(k)) - s_H(k) H(\mathrm{SOC}(k)) \end{aligned} \tag{5.49}$$

where $s_{\mathrm{dis}} = 1$, $s_{\mathrm{chg}} = 0$ when $I(k) - I_f(k) \geq 0$ and $s_{\mathrm{dis}} = 0$, $s_{\mathrm{chg}} = 1$ when $I(k) - I_f(k) < 0$.

5.2.5.2 Extended Kalman Filter Algorithm

Unlike the linear *RC* model used in simulations, the earlier modified ESC model is a nonlinear model that cannot be applied in the linear H_∞ filter mentioned earlier. Therefore, in experiments and applications, we utilize the extended Kalman filter (EKF) to implement the closed-loop component in the proposed framework.

Consider a discrete-time nonlinear system with state vector x_k, input vector u_k, and observation vector y_k at the time instance

k. The dynamic state space model can be represented in the form

$$\begin{aligned} x_{k+1} &= f\left(x_k, u_k, w_k\right) \\ y_k &= g(x_k, v_k) \end{aligned} \tag{5.50}$$

where $f(\cdot)$ is the state evolution function and $g(\cdot)$ is the observation function. Here w_k and v_k stand for process and measurement noise vectors, respectively, with associated covariance matrices $Q = E\left\{w_k w_k^T\right\}$ and $R = E\left\{v_k v_k^T\right\}$. To iterate the evolution process, an initial state vector should be estimated ahead, denoted by x_0. Then the expectation of initial estimates of states and the covariance matrix are given by $\bar{x}_0 = E\left\{x_0\right\}$ and $P_0 = E\{(x_0 - \bar{x}_0)(x_0 - \bar{x}_0)^T\}$.

For application of EKF, it is also necessary to derive the linearized state and observation equations with respect to states and noises near a desired reference point $(\hat{x}_k, u_k, \bar{w}_k, \bar{v}_k)$:

$$\begin{aligned} x_{k+1} &\approx f(\hat{x}_k, u_k, \bar{w}_k) + A_k(x_k - \hat{x}_k) + F_k(w_k - \bar{w}_k) \\ y_k &\approx g(\hat{x}_k, \bar{v}_k) + C_k(x_k - \hat{x}_k) + G_k(v_k - \bar{v}_k) \end{aligned} \tag{5.51}$$

where

$$\begin{aligned} A_k &= \frac{\partial f(x, u_k, \bar{w}_k)}{\partial x}|_{x=\hat{x}_k} \\ F_k &= \frac{\partial f(\hat{x}_k, u_k, w)}{\partial w}|_{w=\bar{w}_k} \\ C_k &= \frac{\partial g(x, \bar{v}_k)}{\partial x}|_{x=\hat{x}_k} \\ G_k &= \frac{\partial g(\hat{x}_k, v)}{\partial v}|_{v=\bar{v}_k} \end{aligned} \tag{5.52}$$

Here $\bar{w}_k = E\left\{w_k\right\}$ and $\bar{v}_k = E\left\{v_k\right\}$ are the expectations of the input noise vector and measurement noise vector, respectively, and $\hat{x}_k$ is the estimate of state vector at time instance k.

The Kalman filter has two distinct phases: prediction and update. The first utilizes the state estimate based on the previous information to produce an estimate of states at the current time. In the update phase, measurement information at the current time is applied to refine this prediction and obtain a revised, and hopefully more accurate, result, again for the current time instance.

1. Predict equations:

$$\begin{aligned} \hat{x}_{k|k-1} &= f(\hat{x}_{k-1|k-1}, u_k, \bar{w}_k) \\ P_{k|k-1} &= A_k P_{k-1|k-1} A_k^T + F_k Q_k F_k^T \end{aligned} \tag{5.53}$$

2. Innovation equations:

$$\begin{aligned} e_k &= y_k - g(\hat{x}_{k|k-1}, \bar{v}_k) \\ S_k &= C_k P_{k|k-1} C_k^T + G_k R_k G_k^T \end{aligned} \tag{5.54}$$

3. Update equations:

$$\begin{aligned} K_k &= P_{k|k-1} C_k^T S_k^{-1} \\ \hat{x}_{k|k} &= \hat{x}_{k|k-1} + K_k e_k \\ P_{k|k} &= P_{k|k-1} - K_k C_k P_{k|k-1} \end{aligned} \tag{5.55}$$

where $\hat{x}_{k|k-1}$ and $P_{k|k-1}$ are the prediction output based on past measurement and corresponding error covariance matrix e_k; S_k is the innovation information obtained from current measurement and corresponding innovation covariance matrix $\hat{x}_{k|k}$; and $P_{k|k}$ are the refined estimate results based on the Kalman gain K_k derived from innovation information.

In the modified ESC model, the above procedure is realized by

$$A_k = \begin{bmatrix} -d_2/d_1 & 0 \\ 0 & 1 \end{bmatrix} \tag{5.56}$$

$$F_k = I \tag{5.57}$$

$$C_k = \begin{bmatrix} \left.\dfrac{\partial V(k)}{\partial I_f(k)}\right|_{\text{SOC}(k)} & \left.\dfrac{\partial V(k)}{\partial \text{SOC}(k)}\right|_{I_f(k)} \end{bmatrix} \tag{5.58}$$

$$G_k = 1 \tag{5.59}$$

5.2.5.3 Performance Demonstration

In experiments, the "real" SOC is calculated based on coulombic counting according to measurement of the electric loader and charger, with both discharging and charging efficiencies of 1.

A combined pulse discharge test is conducted based on the electric loader. A discharge cycle consists of 2-C discharge for 60 s and 3-C discharge for 30 s. As shown in Fig. 5.31, the initial SOC is intended to be different from the real value. In the first round, the current sensor with a measurement range from −200 A to 200 A actually has a bias of around +2 A, causing a gradual increase in the estimation error of the coulomb counting method. EKF has the ability to follow the real SOC trajectory based on the voltage feedback. The average SOC estimation error is 6.235 percent, while the maximum error, with the exception of the errors affected by the initial error, is 11.273

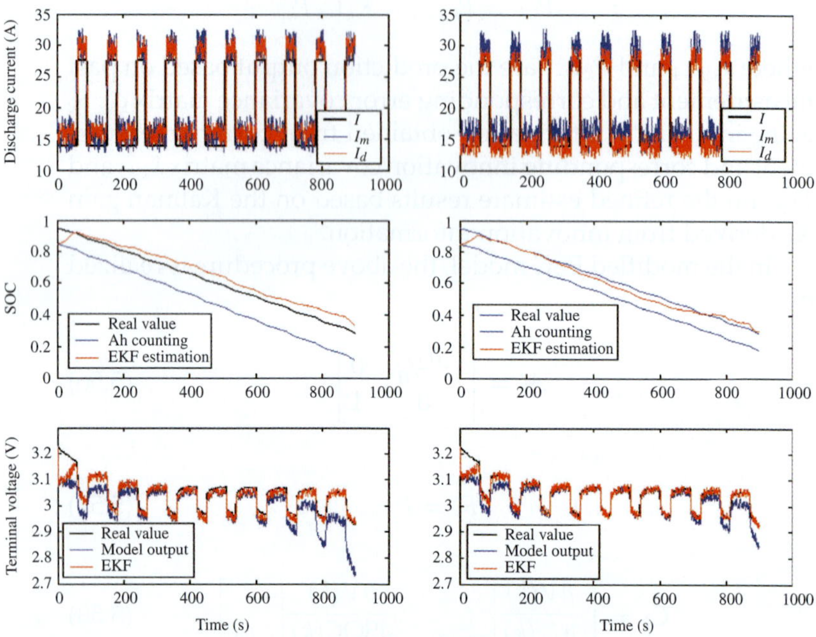

FIGURE 5.31 Discharge experiment results based on electric loader.

percent. In the second round, based on the error of the coulomb counting method, the zero drift of current measurement is estimated as 1.908 A, without the effect of the initial SOC error. By removing the zero drift, estimation results are greatly improved, with an average error of 4.673 percent and maximum error of 7.632 percent.

In the same manner, a charge test is conducted using a combined pulse with 2 C for 60 s and 3 C for 30 s. As shown in Fig. 5.32, in the first round we also assume the zero drift is unknown. The average error is 5.863 percent and the maximum is 9.736 percent. In the second round, the estimated zero drift is 1.836 A, the average error is 4.352 percent, and the maximum error is 7.018 percent.

Since the above results verify that the model based on EKF estimation outperforms the coulombic counting method in every process, the data fusion unit in the real application is designed to trust the results of EKF, and the estimation results

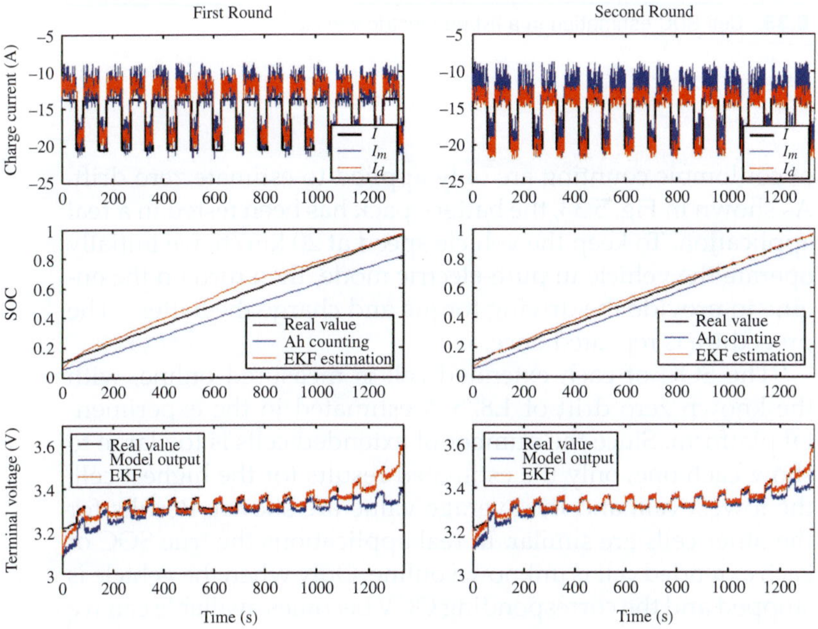

FIGURE 5.32 Charge experiment results based on electric charger.

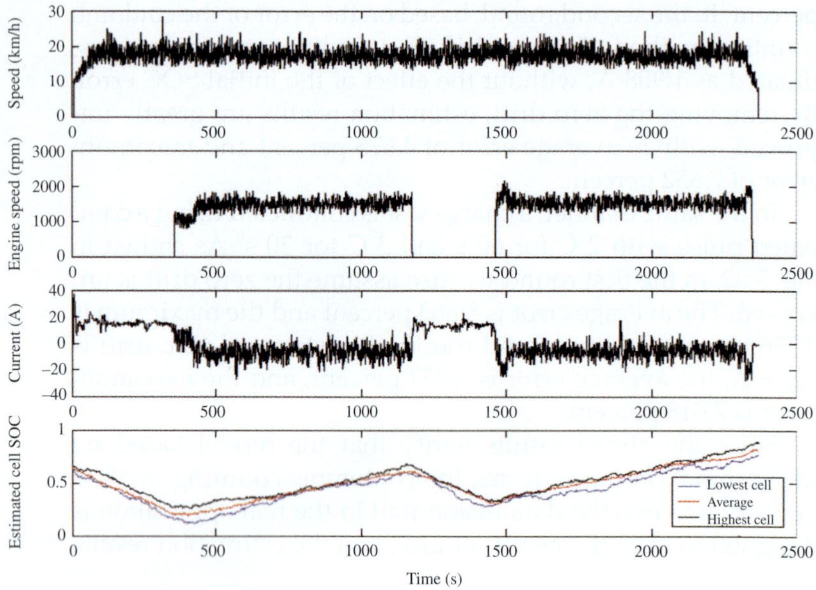

FIGURE 5.33 Cell SOC estimation in a hybrid electric vehicle.

of coulombic counting are only applied to estimate zero drift. As shown in Fig. 5.33, the battery pack has been tested in a real application. To keep the vehicle speed at 20 km/h, we initially operate the vehicle in pure-electric mode, then turn on the engine to provide the driving torque and charge the battery. The procedure is repeated once.

The SOC of each extended cell is measured online, with the known zero drift of 1.875 A estimated in the experimental platform. Since the number of extended cells is too large to show each one, only the estimated results for the highest cell, the lowest cell, and the average value are shown. Results for the other cells are similar. In real applications the true SOC of each extended cell is unknown online. Only when the vehicle is stopped and the corresponding OCV becomes available can we get the "real" value of the final SOC. The maximum SOC estimation error at the end of the running process is 7.635 percent, the minimum is 3.263 percent, and the average is 5.893 percent.

The estimation results satisfy the requirement of 8 percent in China's newly published development goal.[71]

5.2.6 Data Fusion and Overall Performance

From the results shown in Figs. 5.22 and 5.25, it is clear that coulomb counting estimation is generally good at the beginning of the period due to small accumulated errors, while the H_∞ filter obtains better performance in the latter period, when it reaches convergent estimation. Therefore, one natural way to make best use of the advantages and avoid the disadvantages is to combine the results of the two estimators, weighted by time.

In the proposed framework, we design a simple data fusion unit to achieve the final SOC estimation $\widehat{\mathrm{SOC}}(k)$ based on a linear combination of $\widehat{\mathrm{SOC}}_1(k)$ and $\widehat{\mathrm{SOC}}_2(k)$. The weights of the two estimators are experimentally determined by the following equations:

$$\widehat{\mathrm{SOC}}(k) = \begin{cases} \widehat{\mathrm{SOC}}_1(k) & t \leq 5 \\ \omega_1 \times \widehat{\mathrm{SOC}}_1(k) + \omega_1 \times \widehat{\mathrm{SOC}}_2(k) & 5 < t < 10 \\ \widehat{\mathrm{SOC}}_2(k) & t \geq 10 \end{cases} \tag{5.60}$$

where $\omega_1 = -0.2t + 2$, $\omega_2 = 0.2t - 1$, and t is the vehicle running time in minutes.

Since we have demonstrated that coulomb counting with self-learning strategy outperforms the conventional method and the H_∞ filter achieves a more robust estimation than the Kalman filter in a real vehicle driving environment, verification of the availability and efficacy of the overall framework only requires comparison of its performance with the two single estimators.

The simulation environment is the same as described earlier. The SNR of white current noise is fixed at 5 dB, and the zero drift is set to 0.5 A for the first three cycles and changed to 1 A for the remainder. To further test the robustness of the whole framework, we introduce the model error (E_m), defined as $E_m = p_m/p_r$, where p_m is the model parameter, p_r is the real parameter, and p presents the arbitrary parameters of R_o, R_s,

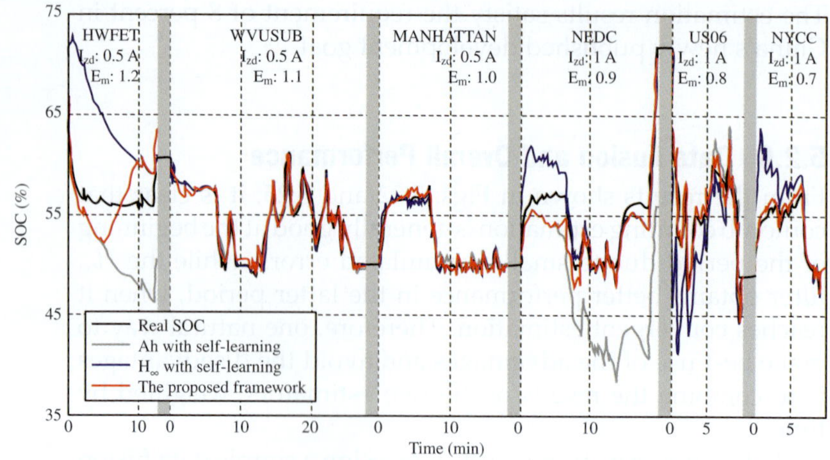

FIGURE 5.34 SOC estimation results of self-learning coulomb counting, the H_∞ filter, and the overall framework.

The gray bars between cycles represent half-hour stops which allow the OCV available. Simulation conditions: white noise with 5-dB SNR, zero drift of 0.5 A for the first through third cycles and 1 A for the fourth through sixth cycles, model error changes from 1.2 to 0.7.

R_b, and C_c. The E_m varies from 1.2 in the first cycle to 0.7 in the last cycle.

The estimation results are shown in Fig. 5.34, and Fig. 5.35 gives the averages of absolute estimation errors and their standard deviations in each cycle.

In the first cycle of HWFET, the zero drift in measurement of current, as well as the model error, results in a large error in the coulomb counting and H_∞ filter. The fusion of the two approaches can reduce this error. In the following two cycles, the self-learning strategy estimates the zero drift and effectively compensates for the error. Although the other two approaches also show good performance, the hybrid approach outperforms them. In the fourth cycle, the drift is increased. The errors of the H_∞ filter and coulomb counting are greatest in the first third and latter two-thirds, respectively, while the hybrid approach has the least error. In the last two cycles, self-learning estimates the zero drift again. Consequently, performance of the Ah method

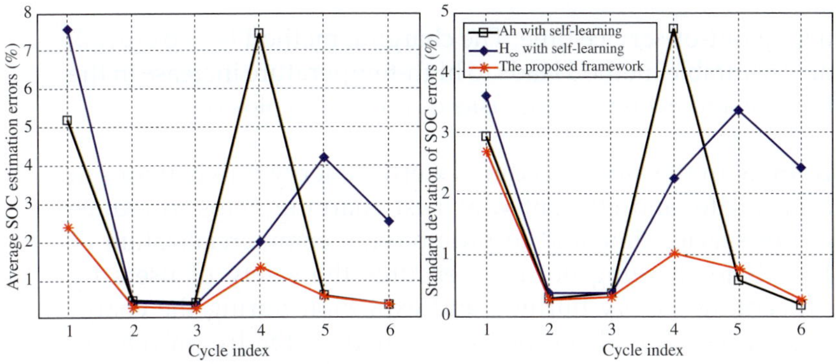

FIGURE 5.35 Numerical indexes of SOC estimation results of self-learning coulomb counting, the H_∞ filter, and the overall framework.

The left figure shows average absolute estimation error while the right figure illustrates their standard deviation. Simulation conditions: white noise with 5-dB SNR, zero drift of 0.5 A for the first through third cycles and 1 A for the fourth through sixth cycles, model error changes from 1.2 to 0.7.

is improved. Even though the model error is increased, due to its robustness the H_∞ filter converges to a small error after passing through a short period of large error, so that the combined results still yield in good performance.

As a summary, the performance of coulomb counting and the H_∞ filter are dependent on the model error, noise, and zero drift. Although their individual performance is slightly better in some regions, the overall performance of the hybrid method is superior overall.

5.3 Fast-Charging Controller

It is necessary to refuel vehicles after a period of travel. For conventional fuel-driven vehicles, the fuel can be replenished fast and safely at filling stations, with an ignorable out-of-service time and without energy loss. However, the replenishment of energy in electric vehicles requires charging the battery pack. In general applications, such as MP3 players, shavers, and cell phones, the charging process lasts for several hours, leading to a long out-of-service time. The charging method is also crucial; an unsuitable method leads to high-temperature increase in the battery and even runs the risks of fire and explosion.

Therefore, charging control is another significant issue in BMS, with the aim of feeding external energy into batteries in a fast, safe, and efficient way. Fast charging helps to reduce out-of-service time and to promote the commercialization of EVs. Safe charging not only ensures the safety of users by preventing battery burning and explosion during the charging process, but also prolongs the battery life by preventing overcharging and overheating damage. Efficient charging can convert as much electric energy as possible from a charger to electrochemical energy stored in a battery to enhance energy efficiency.

5.3.1 Framework Overview

In recent years, much effort has been devoted to establishing battery models, some of which are demonstrated to be accurate and efficient at modeling the behavior of batteries.[4, 42, 84, 96, 103] These models make it possible to apply MPC to management of the battery charging process.

Model predictive control (MPC) is a typical advanced control method and is widely utilized in many fields.[9, 91] It is based on the principle of using a system model to predict system responses to possible future control inputs and to find the best future control sequence to optimize the user-defined objective function. Only the first step of the control sequence is applied each time. The system states are then sampled again, and the calculations are repeated at the next control time. The prediction horizon is constantly shifted forward. Since the receding-horizon strategy updates predictions based on the instant measurement of system inputs and outputs at each control time, its robustness has been demonstrated to be strong.[3] Depending on the linearity of the system models and constraints, MPC may be categorized as either linear MPC or nonlinear MPC. For linear MPC, the best control sequence can be analytically obtained by solving Diophantine equations. However, for nonlinear MPC, the best control sequence can generally be numerically calculated only under an optimization framework.[116]

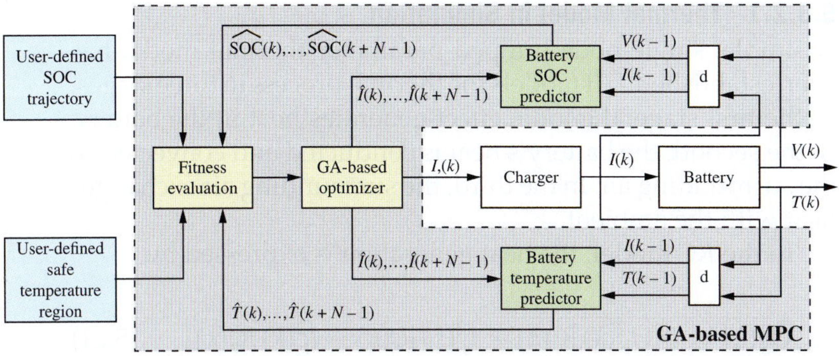

FIGURE 5.36 Fast-charging control framework based on model predictive control.

As shown in Fig. 5.36, the fast-charging control framework proposed in this work consists of the following components:

1. A SOC predictor, predicting the SOC of the battery when fed by a sequence of future charging current
2. A temperature predictor, predicting the future battery temperature under the sequence of charging current
3. A fitness evaluator, evaluating the performance of the sequence of charging current
4. An optimizer, finding the best sequence of charging current using a genetic algorithm (GA).

5.3.2 Predictive Models

To utilize predictive control, battery models are needed to predict future states exerted by a sequence of future charging current. The relevant battery states in this study are SOC and temperature. The former indicates the charging speed and whether overcharge damage is possible, while the latter represents the safety of the charging process and the amount of lost energy converted to heat. Charging control is necessary to (1) maximize charging speed, (2) minimize temperature increases, and (3) restrict temperature to a safe range.

The *RC* model and modified ESC model both have the ability to predict the battery's SOC based on a given future current. Therefore, we directly apply them in the simulation and experiment.

5.3.2.1 Thermal Model in Simulation

A simple, single-node lumped-parameter thermal model has been established.[82] It models the thermal process in three stages. In the first stage, the Joule effect generates heat in the battery. In the second, the battery's heat is conducted and convected to the surrounding air. In the third, the surrounding air exchanges heat with the ambient.

In the *RC* model, the heat generation is expressed by

$$Q_g(k) = T_s\left[I_o^2(k)R_o(k) + I_s^2(k)R_s(k) + I_b^2(k)R_b(k)\right] \tag{5.61}$$

where

$$\begin{aligned} I_o(k) &= I(k) \\ I_b(k) &= \frac{V_b(k) - V_s(k) + R_s(k)I(k)}{R_s(k) + R_b(k)} \\ I_s(k) &= I_b(k) - I(k) \end{aligned} \tag{5.62}$$

Meanwhile, the heat passing from the battery to the surrounding air is expressed by

$$Q_p(k) = \frac{T(k-1) - T_{\text{air}}(k-1)}{R_{\text{eff}}} \tag{5.63}$$

where the T_{air} is the effective temperature of surrounding air and R_{eff} is the effective thermal resistance.

Based on the assumption that 50 percent of the heat from the battery is spent to warm the air, T_{air} is expressed by

$$T_{\text{air}}(k-1) = T_{\text{amb}} - \frac{0.5 \times Q_p(k-1)}{\dot{m}_{\text{air}}C_{\text{air}}} \tag{5.64}$$

where T_{amb} is the ambient temperature, $Q_p(k-1)$ is the passing heat in previous step, $\dot{m}_{\text{air}}$ is the airflow rate, and C_{air} is its heat capacity.

Strictly speaking, R_{eff} will vary depending on the thermal control method; e.g., it will become smaller if the cooling fans are open. However, in this study, we assume the charging process is conducted in an environment with only natural convection and no thermal control device. In this case, R_{eff} is fixed as a constant with the value calculated by

$$R_{\text{eff}} = \frac{1}{hA} + \frac{t}{kA} \tag{5.65}$$

where h is the heat-transfer coefficient in the natural environment, A is the total module surface area exposed to the air, t is the thickness of the module case, and k is the thermal conductivity of the module case material.

Here $Q_g(k)$ indicates the heat generation which heats the battery, while $Q_p(k)$ represents the heat loss which cools the battery by heating the surrounding air. Therefore, the battery temperature $T(k)$ can be calculated by

$$T(k) = T(k-1) + \frac{Q_g(k) - Q_p(k)}{m_{\text{bat}} C_{\text{bat}}} \tag{5.66}$$

where m_{bat} is the battery mass and C_{bat} is its heat capacity.

5.3.2.2 Thermal Model in Experiment

The simulation model requires many physical parameters for the battery and working environment. However, in practice, these parameters cannot usually be obtained accurately. Meanwhile, the theoretical model considers only the ideal situation, which may not be suitable for the complex and nonlinear electrochemical process inside the battery. Furthermore, heat generation is only caused by heating of the resistors, which does not fully represent the actual situation. In reality, heat generation during charge is also affected by the charging acceptance rate. Notten, Veld, and Beek[70] pointed out that "close-to-fully discharged batteries can be recharged with very high currents for a short period of time," and Chow et al.[14] concluded that for A123, a kind of $LiFePO_4$, "at a low state of charge, nearly all the charging current is absorbed by the chemical reaction. Above 80 percent of SOC, more and more energy goes into heat."

Figure 5.37 shows the temperature curves of the experimental cell during charge processes with different charging rates.

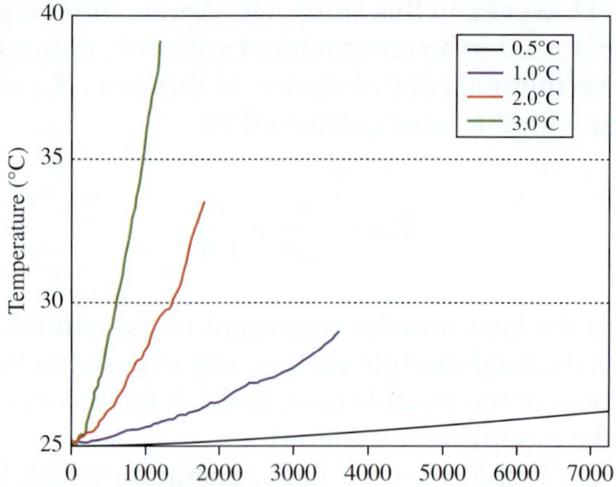

FIGURE 5.37 Temperature rising during various charging rates.

The increasing-temperature trend agrees with the two results reported in the literature earlier. Another two extended cells are also tested, resulting in similar curves.

Besides the SOC, the increasing temperature obviously has a close relationship with the charging rate. High charging rates will cause high temperature rises. Therefore, in practice, we select the SOC, charging rate, and current temperature to predict the future temperature, based on a neural network (NN) trained by the data from three extended cells.

Figure 5.38 shows the structure of the NN implemented to predict temperature. The training method is back-propagation (BP) learning. When the time step is set to 30 s, the average prediction error is 3.36 percent and the maximum is 5.43 percent.

5.3.2.3 Model-Based Prediction

In addition to the models above, to predict battery states under a given sequence of future charging current, two problems still need to be solved: (1) how to initialize the prediction at time k and (2) how to realize the multistep prediction using a one-step predictive model.

To initialize the prediction at time k, the RC model requires the last system state $x(k-1)$, last input $u(k-1)$, and the model

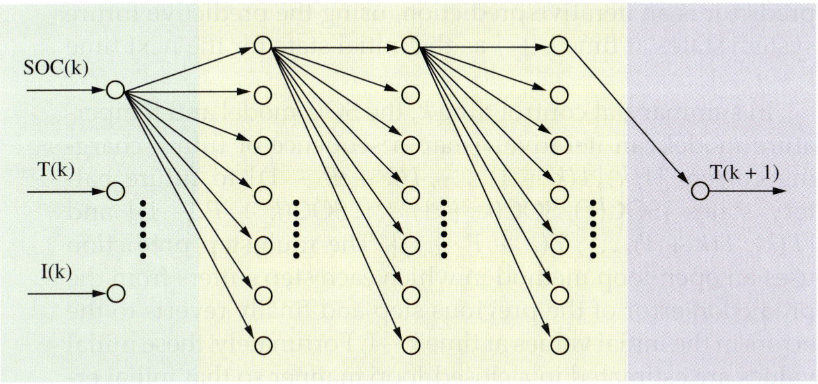

FIGURE 5.38 Neural network for predicting battery temperature.

parameters at time k. Since $u(k-1)$ and $y(k-1)$ are both available from the current sensor and voltage measurement, respectively, $x(k-1)$ can be estimated by closed-loop estimators such as the extended Kalman filter,[85] sigma-point Kalman filter,[86, 87] H_∞ filter,[117] and the robust SOC estimation framework proposed in Sec. 5.2. The closed-loop filters are able to eliminate accumulated errors, and the estimation result will gradually converge to the real value, or to a limited error range. The model parameters are stored in a look-up table indexed by SOC and temperature. The slow change in both SOC and temperature is the reason for updating the model parameters at time k according to SOC$(k-1)$ and battery temperature $T(k-1)$, especially when the T_s is comparatively small. SOC$(k-1)$ is estimated by the method proposed in Sec. 5.2, and $T(k-1)$ is measured directly from a temperature sensor fixed on the surface of the battery.

The above preparation allows a one-step prediction of SOC(k) and $T(k)$, given the value of $I(k)$. However, a single-step prediction is usually insufficient to predict system behavior over a long process. Generally, the prediction horizon in MPC is more than one step. Therefore, a multistep predictor is necessary to predict system states exerted by a sequence of future charging current. A simple way to realize the multistep predictor is an iterative prediction, using the predictive future system states at time $k + j$ as the initial state for the next time $k + j + 1$.

In summary, at control time k, the SOC model and temperature model can iteratively map a sequence of future charging current $[I(k), I(k+1), \ldots, I(k+P-1)]$ to future battery states $[\mathrm{SOC}(k), \mathrm{SOC}(k+1), \ldots, \mathrm{SOC}(k+P-1)]$ and $[T(k), T(k+1), \ldots, T(k+P-1)]$. The multistep prediction uses an open-loop method in which each step suffers from the prediction error of the previous step and finally reverts to the errors in the initial values at time $k-1$. Fortunately, these initial values are estimated in a closed-loop manner so that initial errors are limited to be small. In reality, any battery management system requires such an estimator to obtain the real-time SOC and measure the battery temperature. The results can be used as the initial values.

5.3.3 Formulation under Model Predictive Control Framework

Given a series of control inputs, the length of which is denoted as the prediction horizon P, the future system states can be predicted based on a dynamic system model. The future system behavior under the sequence of inputs can then be evaluated based on a performance index. At each control time k, the basic idea of MPC is to find an optimal sequence of control inputs $[\hat{u}(k), \hat{u}(k+1), \ldots, \hat{u}(k+P-1)]$, which optimizes the performance index, and to apply the first element of the input sequence $\hat{u}(k)$ to the system as the current control variable.

In the charging control problem, MPC optimizes a sequence of future charging current $[\hat{I}(k), \hat{I}(k+1), \ldots, \hat{I}(k+P-1)]$, which has the best performance index based on the predicted battery states $[\widehat{\mathrm{SOC}}(k), \widehat{\mathrm{SOC}}(k+1), \ldots, \widehat{\mathrm{SOC}}(k+P-1)]$ and $[\hat{T}(k), \hat{T}(k+1), \ldots, \hat{T}(k+P-1)]$.

5.3.3.1 Performance Indexes

The performance index reflects the control objectives. The first objective is to maximize charging speed. However, since it is impossible to directly predict the total charging duration in a limited prediction horizon, we turn to the tracking of a user-defined SOC trajectory instead. A fast-rising SOC trajectory requires a fast charging speed while a flat one requires a slow speed.

The expected SOC trajectory can directly copy from any real charging trajectory controlled by any charging scheme. In addition, the expected trajectory can be set as a real trajectory with revisions based on certain special considerations. For example, for a constant-current, constant-voltage (CCCV) trajectory, we may increase the charging speed when the SOC is in the middle and the internal resistance is small. In addition, the expected trajectory can be designed entirely on the demands of the charging system.

To track the expected trajectory, a part of the performance index J_1 is expressed by

$$J_1 = |\widehat{\mathrm{SOC}}(k + P - 1) - \mathrm{SOC}^*(k + P - 1)| \tag{5.67}$$

where SOC* is the expected SOC trajectory. Evaluation of the SOC tracking performance is based on only the final prediction state because the SOC only needs to achieve the expected point in the final step. How this is achieved is not very important from the point of view of the charging process.

The second objective of charging control is to minimize the increase in temperature, which partially reflects energy efficiency and system safety. Assuming that the expected trajectory is copied from a CCCV charging process, without consideration of temperature, the best charging current sequence is the same sequence applied in the CCCV process. However, if we include evaluation of performance based on temperature, as expressed in J_2, the MPC will try to achieve the expected SOC point in the final step, by the process in which temperature rise is minimized.

$$J_2 = \max\{\hat{T}(k + j) - T(k - 1) | j = 0, 1, \ldots, P - 1\} \tag{5.68}$$

5.3.3.2 Constraints

Constraints in the charging process are designed to prevent a charging sequence that may cause damage to the battery or lead to dangerous events. The first constraint is that the SOC must not exceed 100 percent, to avoid overcharge damage. In practice, to reserve some tolerance, 98 percent SOC is treated as the full charging state. The second constraint is that temperature must be kept in a user-defined range to avoid overheating caused by overcharge or by a large charging current that exceeds the instant charging acceptance level. The two constraints are expressed by

$$C_1 : \widehat{\text{SOC}}(k + j) \leq 98\% \tag{5.69}$$

and

$$C_2 : \hat{T}(k + j) \leq \bar{T}(k + j) \tag{5.70}$$

where $j = 0, 1, \ldots, P - 1$ and $\bar{T}$ is the user-defined safe range, which is designed either as a constant, indicating the highest temperature during the whole charging process, or as a time-indexed function specifying the temperature limitation along with charging duration.

5.3.4 Optimization Using Genetic Algorithm

The MPC charging control problem is formulated to minimize the performance indexes J_1 and J_2, subject to the constraints C_1 and C_2. Essentially, the control problem is transformed into a constrained multiobjective optimization problem.

Generally, a multiobjective optimization problem can be solved either by multiobjective optimizers directly or by first transforming it into a single-objective problem that is then solved by single-objective optimizers.[17] In this work, we apply the latter method because SOC tracking is more important than temperature management in the charging process, so the two objectives can be merged into one index by summing with different weights. To solve the optimization problem, a genetic algorithm (GA) is applied due to its strong global searching ability, without the requirement for derivative information of the objective function. Since the two constraints should be strictly satisfied in the charging process to ensure the safety and health of the battery, solutions that fail to satisfy any one of the constraints will be assigned the worst fitness.

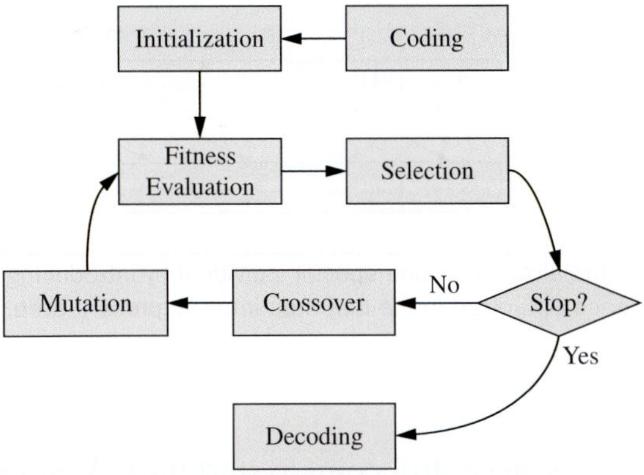

FIGURE 5.39 Scheme of the standard GA.

Based on the earlier discussion, the fitness function to be minimized in GA is expressed by

$$F = \begin{cases} \omega_1 J_1 + \omega_2 J_2 & C_1 \text{ and } C_2 \text{ are held} \\ +\infty & \text{otherwise} \end{cases} \tag{5.71}$$

where ω_1 and ω_2 are the weights of J_1 and J_2, respectively.

The minimization problem is solved by a standard GA. the scheme of which is illustrated in Fig. 5.39 and briefly described in the following steps.

Coding The standard GA generally codes a candidate solution as a string of characters which are usually binary digits, referred to as a *chromosome*. The candidate solution is termed an *individual*. Accordingly, the set made up of a number of individuals is termed a *population*. In this chapter, we apply a real-value coding method, which codes a candidate solution as a set of floating-paint decision variables. The real-value coding method is proved to have superior performance to the binary-coded method in control optimization problems.[41]

Initialization The standard GA starts with the generation of an initial population. Usually, individuals in the initial population are produced randomly. In MPC, the initialization process

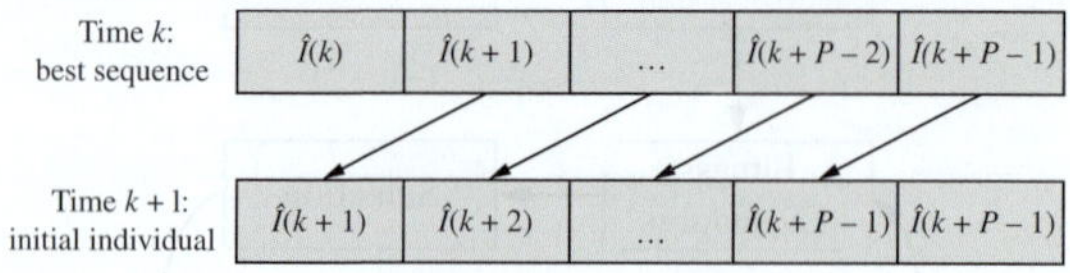

FIGURE 5.40 Initialization of one special individual by introducing the best control sequence optimized in the last step into the present step.

is executed at each control point to start the GA. Since the best control sequence optimized at time k contains good candidates from $k+1$ to $k+P-1$, as shown in Fig. 5.40, one initial individual is specially designed by shifting it one time step and filling the last charging current with the same value as $\hat{I}(k+P-1)$. This individual introduces historical best charging sequence into the current optimization process; thus it is helpful to improve optimization performance to be at least very similar to the previous prediction.

Fitness Evaluation We can evaluate the fitness of each individual in each generation according to Eq. (5.71). The smaller the fitness, the better the individual. However, to facilitate the following selection steps, the raw fitness is usually scaled to assign suitable selection pressure to each individual. In this work, the scale function is expressed as

$$F_{\text{scale}} = \frac{1}{1+F^r} \tag{5.72}$$

where r is power of raw fitness. A large r will quickly increase the selection pressure to a worse individual, accelerate the convergence speed, and increase the risk of premature battery failures especially for multipeak landscape, and vice versa.

Selection Individuals are selected from the last generation to the next generation based on the scaled fitness F_{scale} following the *survival of the fittest* rule. Many selection methods have been developed to avoid genetic drift and premature phenomena.

In this work, the roulette wheel selection method is adopted.[29] The elitism strategy is also applied in selection to ensure that the best solution will never be lost.

Crossover In the crossover step, the standard GA exchanges information between two parent individuals and produces two child individuals. In this work, the arithmetical crossover method is used. Given two parents x_1 and x_2, the children y_1 and y_2 are produced by linear combinations of parents with a random coefficient $\lambda \in R_\lambda$:

$$\begin{aligned} y_1 &= \lambda x_1 + (1-\lambda)x_2 \\ y_2 &= \lambda x_2 + (1-\lambda)x_1 \end{aligned} \tag{5.73}$$

Mutation After the crossover step, a subset of individuals is selected with a mutation probability of P_m. To explore the search space, we use gaussian mutation, which adds a random value from a gaussian distribution with variance σ to each item of the selected individual.

Termination Many terminating conditions have been proposed to stop the iteration process, for example, when the distances among individuals are smaller than a predetermined value, an individual satisfies a minimum criterion, or the maximum number of generations is reached. The last method is applied here.

5.3.5 Performance Demonstration

5.3.5.1 Settings

To evaluate the proposed MPC-based charging control strategy, simulation experiments are conducted based on a well-established "7 Ah Saft lithium ion battery" provided in Advisor. *RC* model parameters and OCV all are time-variant variables depending on SOC and temperature. The curves of OCV and R_o are shown in Fig. 5.41 as examples. The interpolating method is applied to create the continuous values space. The constant parameters in this simulation environment are given in Table 5.4.

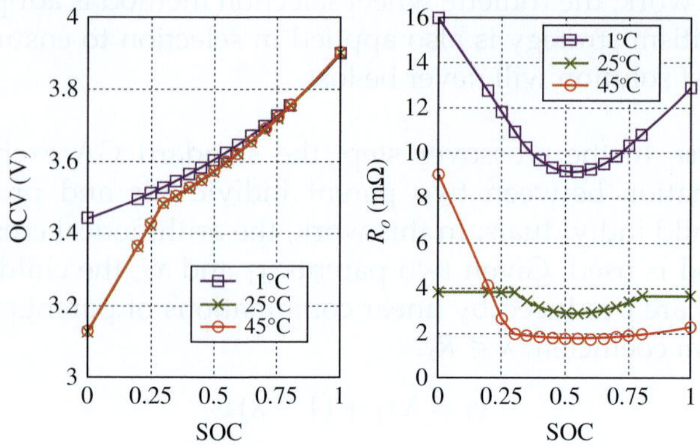

FIGURE 5.41 Battery time-variant properties, taking OCV and R_o as examples. Data source: Advisor.

	Symbol	Description	Value	Unit
Battery (simulation)	C	Battery nominal capacity	7	Ah
	m_{bat}	Battery mass	0.37824	kg
	C_{bat}	Battery heat capacity	795	J/kg · K
	R_{eff}	Effective thermal resistance	7.8146	K/W
	T_{amb}	Ambient temperature	20	°C
	$\dot{m}_{air}$	Airflow rate	5.8333	g/s
	C_{air}	Air heat capacity	1009	J/kg · K
MPC	T_s	Control period	30	s
	P	Prediction horizon	5	—
	ω_1	Weight of SOC tracking J_1	100	—
	ω_2	Weight of temperature rising J_2	1	—
GA	MaxGen	Maximum generation number	30	—
	PopSize	Population size	50	—
	r	Power of raw fitness in scaling	2	—
	R_λ	range of crossover coefficient	[0.1, 0.9]	—
	P_m	Mutation probability	0.2	—
	σ	Variance of gaussian mutation	1	—

TABLE 5.4 Simulation and Experiment Parameters Setting

Remark: The typical value of J_1 is around 0.08 while that of J_2 is around 2. Therefore, the real weight ratio of J_1 to J_2 is around 4:1.

In this work, we select the control period T_s to be 30 s, using a trial-and-error method according to experimental results. The shorter period requires the charger to change the charging current more frequently. It will increase the working load to the charger and lead to greater energy loss caused by switch circuits. The longer period will decrease the frequency to tune the charging current in the charging process so that the performance of MPC will be limited. In addition, the long control period implies that the prediction based on the model is over a long time, which essentially requires a more accurate battery model. Finally, the 30-s control period is long enough to allow GA to finish the optimization process.

5.3.5.2 Evaluation Method

Fast charging speed and low heat generation are both objectives in charging control. However, the two goals conflict with each other. Fast charging essentially requires large current and hence leads to high heat generation, and vice versa. From the viewpoint of multiobjective optimization, the conflicting objectives are usually evaluated by the Pareto curve. For charging control, the x axis is set as the charging duration and the y axis as the final temperature increase $T(\text{end}) - T(0)$. As shown in Fig. 5.42, the curve with the circle marks represents the Pareto front of the CCCV family, where the CC period applies 1.5-C to 6-C current for fast charging. The CCCV Pareto front splits the objective space into two sections. Any charging result located in the upper right section is worse than the CCCV family for both objectives, while any result in the bottom left section is better than CCCV for both objectives. Therefore, we evaluate the performance of the charging controller according to the location of results.

5.3.5.3 Simulation Results

To facilitate comparison with the conventional CCCV family, we set the expected SOC trajectory by multiplying by 1.05. The new trajectory is intended to accelerate the charging speed. Since the working temperature of Li-ion batteries, especially for vehicular batteries, is from 20 to 40°C, the ambient temperature is set as 20°C, and the safe range of temperature is set below

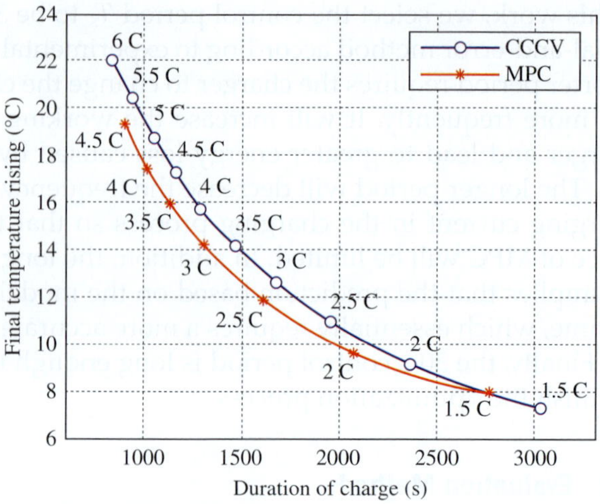

FIGURE 5.42 Pareto fronts of CCCV and MPC charging methods in simulation. The expected trajectories of MPC are modified from CCCV by multiplying by 1.05.

40°C. The search space of the charging current is fixed from 0 to 6 C.

The MPC is applied to track the modified SOC trajectories from 1.5 to 4.5 C. A faster charging speed cannot keep the rising temperature in the safe range. The Pareto front of the MPC is illustrated in Fig. 5.42 by the curve marked with stars. The results for the MPC clearly dominate those of the CCCV family. Only the result of the trajectory revised from 1.5 C has a similar performance to CCCV. The reason is that a 1.5-C current is comparatively so small that it limits the applicable current sequence to a small range. If the fixed search range can be accordingly reduced, e.g., from 0 to 3 C, the result will be improved.

As an example, the charging processes of CCCV and MPC are compared in the case of 3 C, shown in Fig. 5.43. At the beginning of charge, the internal resistance is large when the SOC is very low. In this process, the rising temperature dominates the fitness function. The optimized charging profile applies a smaller current than CCCV. However, the increase in the SOC tracking error gradually requires a higher current

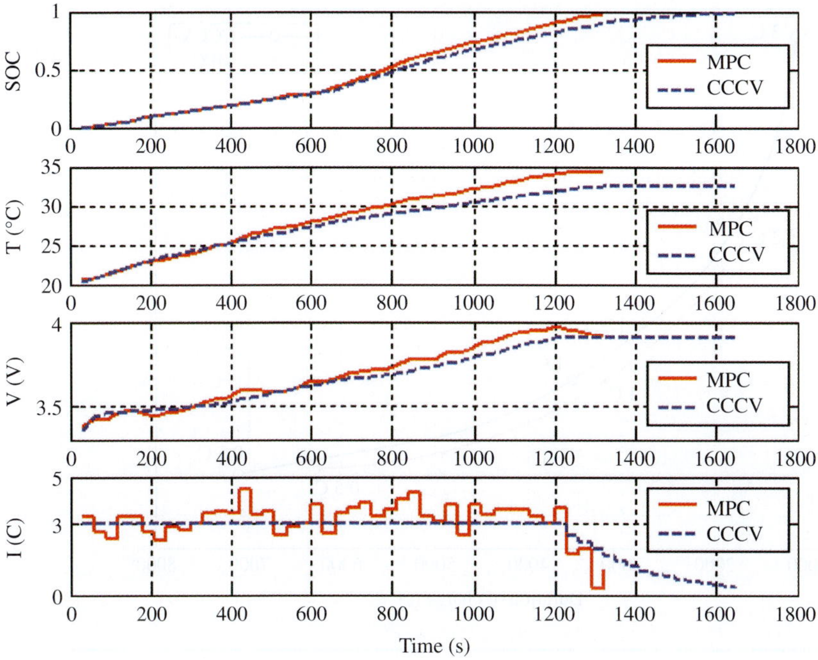

FIGURE 5.43 Curves during CCCV and MPC charging processes using modified 3-C profile.

to speed up. Meanwhile, internal resistance is reduced significantly along with the increase in SOC. Therefore, in the middle period, the charging current for MPC is larger than for CCCV. In the final stage, the current in the CV period is decreased to prevent overvoltage of the terminal voltage. Since the trajectory of the MPC is modified from the CCCV, the same trend is retained in MPC, keeping the terminal voltage under 4 V.

5.3.5.4 Experiment Results

Since the maximum charging current of the programmable electric charger is 9 A, in the experiment we use a single cell with 2.3 Ah to demonstrate the fast-charge performance of MPC. Correspondingly, the charge resistance model of a 6.9-Ah cell is multiplied by 3. To protect the cells, we limit the charging

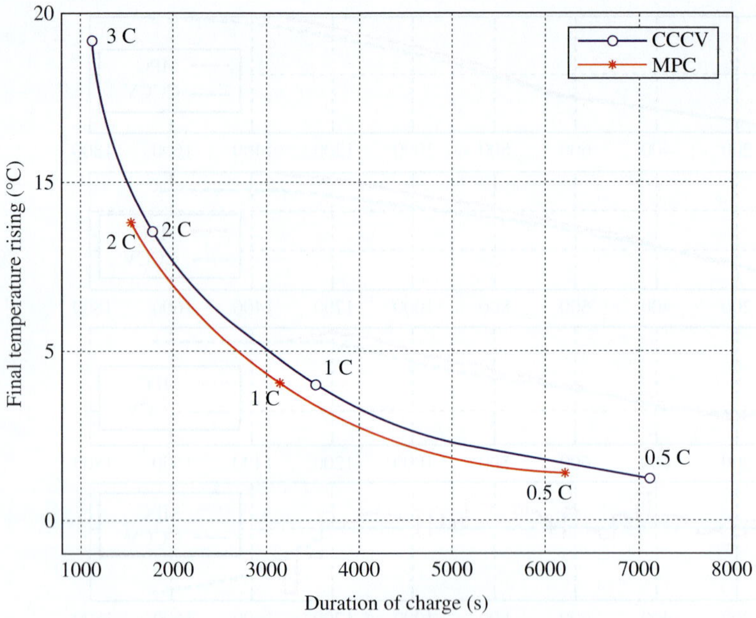

FIGURE 5.44 Pareto fronts of CCCV and MPC charging methods in experiments. The expected trajectories of MPC are modified from CCCV by multiplying by 1.10.

current to below 3 C, that is, 6.9 A. The ambient temperature is 25°C. The expected SOC trajectories of MPC are 1.10 times those of the CCCV results.

Figure 5.44 shows the Pareto fronts of the CCCV and MPC charging results in experiments. All the results for MPC are superior to those for CCCV. Figure 5.45 illustrates the details of the 2-C charging processes. Unlike the simulation model, where the temperature is high in the lower SOC range of the specified battery type, the real $LiFePO_4$ cell has good charging acceptance at low SOC. Therefore, a high charging current is utilized to increase the charging speed. Before reaching around 80 percent SOC, another high-charging period occurs. This is so because MPC predicts the high temperature increase beyond 80 percent, so it applies a high charging current at this stage and switches to a low charging current when SOC is close to 1.

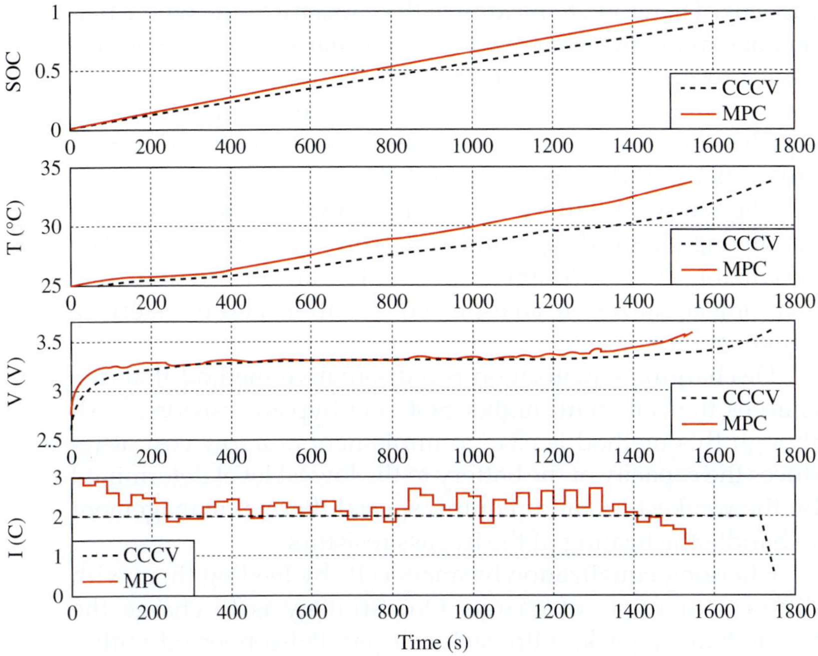

FIGURE 5.45 Experimental curves during CCCV and MPC charging processes using modified 2-C profile.

5.4 Battery Equalization

The existing technologies in battery design and production cannot retain the consistency of cell parameters, such as internal resistance, charge and discharge efficiencies, speed of aging and degradation, and so forth. The differences will result in the imbalance of cells in the same battery package and will force cells to have different states of charge and terminal voltages. Therefore, monitoring and control at the battery level are insufficient to manage a battery, especially a vehicular battery.

The most significant problem caused by cell imbalance is the SOC difference among cells. The battery behavior is a typical cask effect, where the cell with the lowest SOC determines the discharge capacity and the highest determines the charge capacity. Therefore, to maximize the capacity of the whole battery and avoid overcharge and overdischarge damage, it is necessary to keep the SOC differences within a tolerably small range. This function, known as battery equalization or cell equalization, has become one of the most difficult and important issues in BMS, especially for EVs.

The study of battery equalization mainly focuses on the equalizing circuit design and equalizing current control. The equalizing circuit design is generally categorized into discharging equalization, charging equalization, and transferring equalization.

Discharging equalization is a dissipative method that discharges the cells with higher SOC via bypass resistors.[57] Although this method is easy to implement at a low cost, it reduces the capacity of the battery to the lowest level determined by the weakest cell and requires special thermal management to handle the heating of the bypass resistors.

Charging equalization balances cells by feeding them with different amounts of charge. One strategy is to charge the whole battery pack, with each cell parallel-connected with a shunt resistor. A MOSFET series connected with each shunt resistor is controlled to tune the shunt current.[66] A simple design based on this principle directly disconnects the fully charged cells from the battery pack.[23] Another way is to route a boost current to the weaker cells through a selective relay matrix.[31, 120] The disadvantage of charging equalization is that it works only when used with a charger; thus it cannot work in hybrid electric vehicles and during the vehicle running process.

Transferring equalization is an advanced equalization method that delivers the charge from higher-SOC cells to lower ones to achieve a balance. This method results in little energy loss and can work without any external components, thus it has attracted a lot of research interest in recent years. The charge transfer can be implemented by a capacitor, inductor, and transformer. Capacitor- or inductor-based transfer is usually utilized in cell-to-cell equalization. The higher-voltage cell first discharges the electric energy to be stored in the capacitor or inductor, which then releases the stored energy to charge the lower-voltage cells. The time sequence is controlled by MOSFET according to a pulse-width modulation

(PWM) signal.[52, 54, 55] Transformer-based equalization is often employed in battery pack or module pack cell equalization, in which the selected cells, usually the weakest cells, are charged using the energy of the whole battery pack. The transformer is applied to convert the high voltage of the battery to the low voltage of the cell. The equalizing current is also controlled by a MOSFET according to the PWM signal.[78, 105] Further details on equalizing circuit design are reviewed in Refs. 23, 66, and 77.

Compared with the intensive designs of equalizing circuits, control of the equalizing current has still not been comprehensively studied. The equalizing current determines the duration of equalization and is usually designed to be large when the difference in terminal voltage between the selected or neighbored pair of cells is large, and to be small when the difference is small. It also takes into consideration the terminal voltage—the higher it is, the smaller the current. Based on these general concepts, a fuzzy control system is proposed[53] to tune the equalizing current in real time. However, the state that causes the difference in terminal voltage, and that should really be equalized, is the SOC. Although the open-circuit voltage (OCV) can indirectly represent SOC, a small difference in OCV may be caused by a large difference in SOC, due to the nonlinear and flat voltage-SOC relationship. Therefore, Speltino et al.[101] have recently proposed an equalization control method directly based on SOC estimation; nevertheless, the SOC is only utilized as a threshold to turn the equalization process on and off, rather than to tune the equalizing current.

5.4.1 Equalizing Circuit and Analysis

5.4.1.1 Battery Modeling

To analyze the dynamics of the equalizing circuit, it is necessary to model the cell. One method is to use a capacitor that simulates energy storage and a series-connected resistor that simulates internal resistance.[53] This cell model leads to a linear relationship between the residual charge and OCV, as expressed by $Q = CV$. However, in reality, the SOC-OCV relationship is typically a nonlinear curve, especially for lithium-ion batteries, as shown in Fig. 5.46. The internal resistance is also a

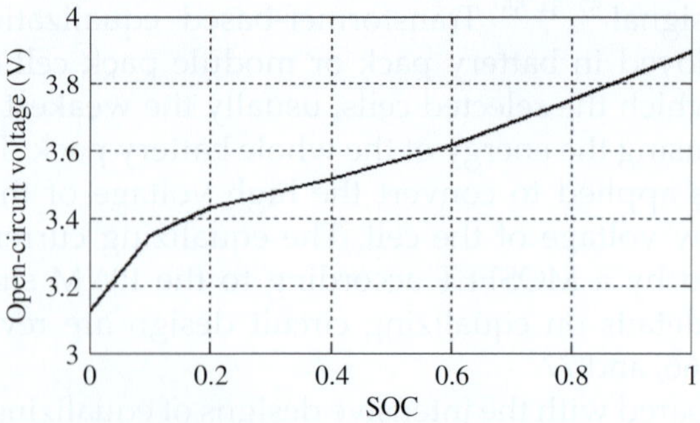

FIGURE 5.46 Typical OCV-SOC curve of a Saft 6-Ah Li-ion cell at 25°C, provided by Advisor.

changeable variable, along with SOC, as shown in Fig. 5.47. Therefore, a capacitor-resistor model is insufficient to model the cell dynamics in battery equalization.

Considering that in a MOSFET control cycle only a very few charges are transferred, the variation in cell voltage will be small enough to be ignored. Thus, we model the cell as a voltage source with a series-connected resistor. The voltage of the voltage source and the resistance of the resistor are both determined by the SOC, according to the cell characteristic curves obtained either by testing or from the manufacturing datasheet.

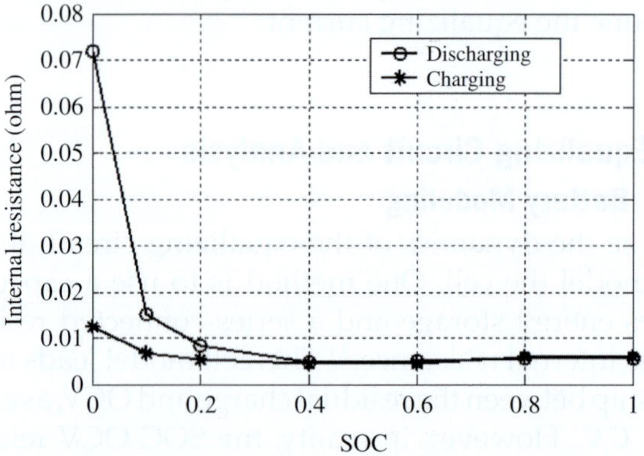

FIGURE 5.47 Typical internal resistance versus SOC curve of a Saft 6-Ah Li-ion cell at 25°C, provided by Advisor.

5.4.1.2 Equalizing Circuit and Principle

As shown in Fig. 5.48, the framework of battery equalization in this section is based on individual cell equalizers (ICEs), which balance each pair of neighboring cells.[53] A battery pack containing *n* cells requires (*n* – 1) ICEs.

Since an ICE works on a cell-to-cell level, only Cell 1 and Cell 2 are used to facilitate the following illustration and analysis, as shown in Fig. 5.49*a*. The implementation of each ICE has multiple options, as reviewed in Sec. 4.1. In this study, we choose the modified bidirectional Cûk converter, as show in Fig. 5.49*b*.

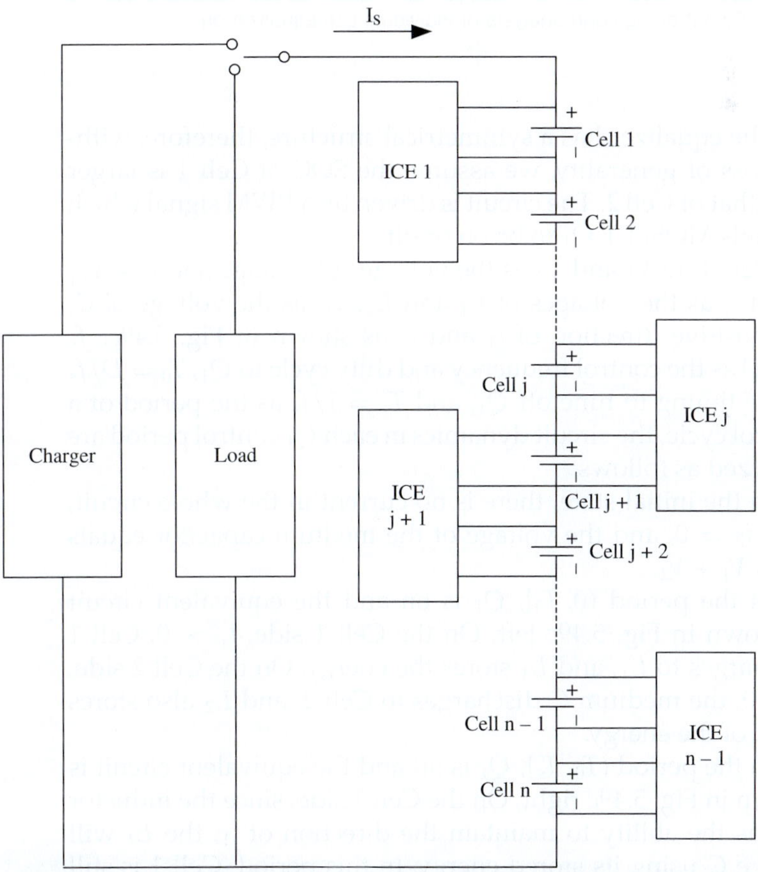

FIGURE 5.48 The framework of individual cell equalization system.

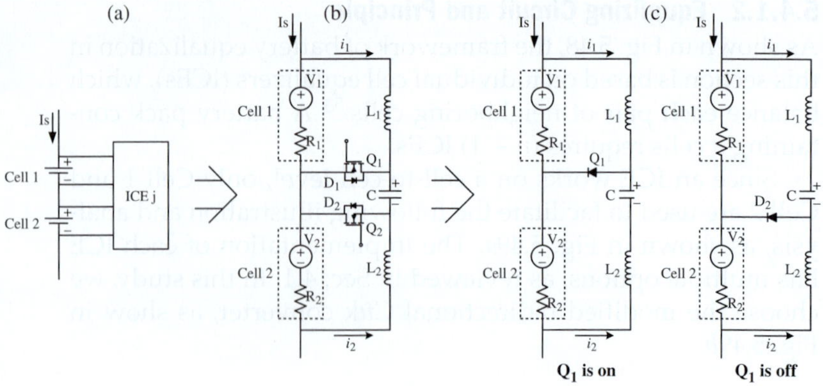

FIGURE 5.49 Circuit design and analysis of individual cell equalization.

The equalizer has a symmetrical structure; therefore, without loss of generality, we assume the SOC of Cell 1 is larger than that of Cell 2. The circuit is driven by a PWM signal which controls MOSFET Q_1 to be on or off.

Denoting V_1 and V_2 as the voltages of voltage sources, V_{L_1} and V_{L_2} as the voltages of L_1 and L_2, V_C as the voltage of C, the positive direction of i_1 and i_2 as shown in Fig. 5.49c, f_s and D as the control frequency and duty cycle to Q_1, $T_0 = D/f_s$ as the timing to tune off Q_1, and $T_s = 1/f_s$ as the period of a control cycle, the circuit dynamics in each Q_1 control period are analyzed as follows:

In the initial state, there is no current in the whole circuit, $i_1 = i_2 = 0$, and the voltage of the medium capacitor equals $V_C = V_1 + V_2$.

In the period $(0, T_0]$, Q_1 is on and the equivalent circuit is shown in Fig. 5.49c left. On the Cell 1 side, $i_1 > 0$, Cell 1 discharges to L_1, and L_1 stores the energy. On the Cell 2 side, $i_2 > 0$, the medium C discharges to Cell 2, and L_2 also stores some of the energy.

In the period $(T_0, T_s]$, Q_1 is off and the equivalent circuit is shown in Fig. 5.49c right. On the Cell 1 side, since the inductor L_1 has the ability to maintain the direction of i_1, the L_1 will charge C using its stored energy. In this period, Cell 1 is still discharging. The D_2 limits the direction of i_1; thus when $i_1 = 0$, the circuit reaches a stable state. In the same manner, on the

Cell 2 side, L_2 also charges continuously Cell 2 until the stored energy is totally transferred. When i_2 reduces to 0, the circuit reaches a steady state.

Through careful selection of L_1, L_2, C, f_s, and D, the final states of L_1, L_2, and C after a Q_1 control cycle can be reset to the same as their initial states. In this case, the equalization behavior will repeat in the following cycles, and the charges are transferred from Cell 1 to Cell 2 gradually. The quantitative analysis of this circuit is presented in the following section.

5.4.1.3 Quantitative Analysis

1. *Governing equations of* i_1: When $i_1 \geq 0$, it is governed by

$$\begin{aligned} &L_1 i_1'(t) + R_1 i_1(t) - V_1 = 0 && t \in [0, T_0] \\ &C L_1 i_1''(t) + C R_1 i_1'(t) + i_1(t) = 0 && t \in (T_0, T_s] \end{aligned} \tag{5.74}$$

Once $i_1 < 0$, it will limited to 0 by D_2. The initial and jumping conditions are given by

$$\begin{aligned} &i_1(0) = 0 \\ &i_1'(0) = V_1/L_1 \\ &i_1'(T_0 + 0^+) = [V_1 - R_1 i_1(T_0) - V_C(T_0)]/L_1 \end{aligned} \tag{5.75}$$

2. *Governing equations of* i_2: When $i_2 \geq 0$, it is governed by

$$\begin{aligned} &C L_2 i_2''(t) + C R_2 i_2'(t) + i_2(t) = 0 && t \in [0, T_0] \\ &L_2 i_2'(t) - R_2 i_2(t) + V_2 = 0 && t \in (T_0, T_s] \end{aligned} \tag{5.76}$$

Once $i_2 < 0$, it will limited to 0 by D_2. The initial and jumping conditions are given by

$$\begin{aligned} &i_2(0) = 0 \\ &i_2'(0) = V_1/L_2 \\ &i_2'(T_0 + 0^+) = [R_2 i_2(T_0) - V_2]/L_2 \end{aligned} \tag{5.77}$$

3. *Governing equations of V_C:*

$$V_C(t) = \begin{cases} V_C(0) - \dfrac{\int_0^t i_2(x)dx}{C} & t \in [0, T_0] \\ V_C(T_0) + \dfrac{\int_{T_0}^t i_1(x)dx}{C} & t \in (T_0, T_s] \end{cases} \tag{5.78}$$

with the initial condition

$$V_C(0) = V_1 + V_2 \tag{5.79}$$

Considering generally $R_1, R_2 \sim 10^{-2}\Omega$, $L_1, L_2 \sim 10^{-4}H$, and $C \sim 10^{-4}F$, the solutions of the above governing equations are given as follows:

1. *Solution of i_1:*

$$i_1(t) = \begin{cases} \dfrac{V_1}{R_1} - \dfrac{V_1}{R_1} e^{-(R_1/L_1)t} & t \in [0, T_0] \\ e^{\alpha_1 t}\left[b_1 \cos(\beta_1 t) + b_2 \sin(\beta_1 t)\right] & t \in (T_0, T_{11}] \\ 0 & t \in (T_{11}, T_s] \end{cases} \tag{5.80}$$

where

$$\begin{aligned} \alpha_1 &= -\frac{R_1}{2L_1} \\ \beta_1 &= \frac{\sqrt{4CL_1 - R_1^2C^2}}{2CL_1} \\ b_1 &= \frac{\beta_1 i_1(T_0)\cos(\beta_1 T_0) - i_1'(T_0)\sin(\beta_1 T_0) + \alpha_1 i_1(T_0)\sin(\beta_1 T_0)}{\beta_1 e^{\alpha_1 T_0}} \\ b_2 &= \frac{\beta_1 i_1(T_0)\sin(\beta_1 T_0) + i_1'(T_0)\cos(\beta_1 T_0) - \alpha_1 i_1(T_0)\cos(\beta_1 T_0)}{\beta_1 e^{\alpha_1 T_0}} \\ T_{11} &= \frac{1}{\beta_1}\arctan\left(\frac{-b_1}{b_2}\right) \end{aligned} \tag{5.81}$$

2. *Solution of i_2:*

$$i_2(t) = \begin{cases} e^{\alpha_2 t}\dfrac{V_1}{\beta_2 L_2}\sin(\beta_2 t) & t \in [0, T_0] \\ \dfrac{V_2}{R_2} + Ae^{(-R_2/L_2)t_0} & t \in (T_0, T_{12}] \\ 0 & t \in (T_{12}, T_s] \end{cases} \quad (5.82)$$

where

$$\begin{aligned} \alpha_2 &= -\frac{R_2}{2L_2} \\ \beta_2 &= \frac{\sqrt{4CL_2 - R_2^2C^2}}{2CL_2} \\ A &= e^{(-R_2/L_2)T_0}\left[i_2(T_0) - \frac{V_2}{R_2}\right] \\ T_{12} &= T_0 + \frac{L_2}{R_2}\ln\frac{-V_2}{R_2 i_2(T_0) - V_2} \end{aligned} \quad (5.83)$$

3. *Necessary conditions:*
The above solutions are subjected to the following necessary conditions. When design a controller, D and f_s must satisfy them.
   a. To ensure Cell 2 is charged in Q_1 on period: $T_0 \leq \pi/\beta_2$
   b. To ensure Cell 1 is still discharged after T_0: $T_{11} > T_0$
   c. To ensure Cell 2 is still charged after T_0: $T_{12} > T_0$
   d. To ensure $i_1(T_s) = 0$: $T_s \geq T_{11}$
   e. To ensure $i_2(T_s) = 0$: $T_s \geq T_{12}$

5.4.1.4 Implementation of an Equalizing Circuit

To implement an equalizing circuit for the mentioned 6-Ah cells, we select equalizing parameters $L_1 = L_2 = 100$ μH, $C = 470$ μF, and $D = 0.45$. Although the internal resistances of cells will change along with their SOCs, in this subsection, 10 mΩ is used as a typical value to facilitate the following discussion.

The control frequency f_s is utilized as the control variable of the equalization system. Given f_s =5 kHz, the typical current and voltage curves in two successive Q_1 control cycles are shown in Fig. 5.50. It is clear that the dynamics of the

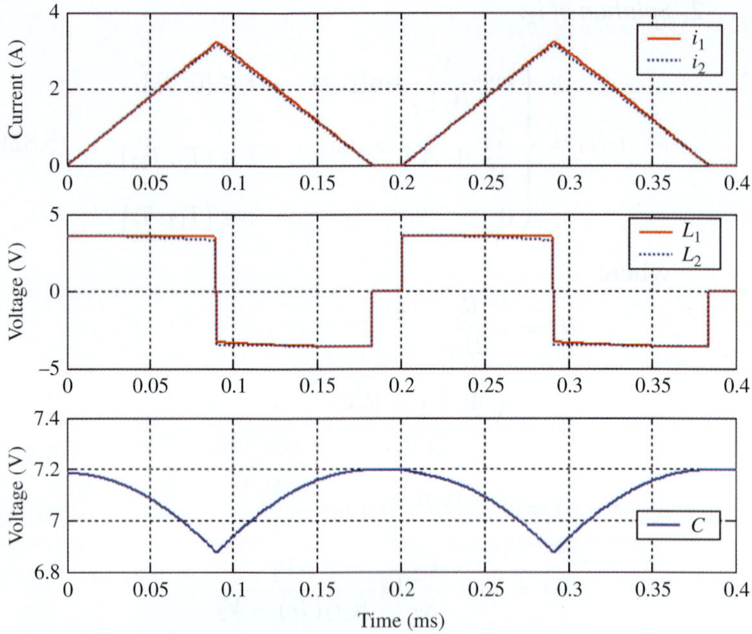

FIGURE 5.50 The current and voltage curves in two successive MOSFET control cycles ($f_s = 5$ kHz, $SOC_1 = 0.6$, $SOC_2 = 0.5$).

equalization system satisfy the theoretical analysis. One remarkable feature of the dynamics is the small difference between i_1 and i_2, which causes the current curves not to overlap completely. The difference is due to the energy lost on the internal resistors of cells. The charge drawn from Cell 1 is always larger than that transferred into Cell 2. Therefore, the ratio of i_2 to i_1 can be used as an indicator of the equalizing efficiency. As shown in Fig. 5.51, the smaller the control frequency, the larger the equalizing current. Hence, the energy loss on internal resistors increases, resulting in lower energy efficiency. In practice, capacitors, inductors, and MOSFETs also cause energy loss, and so energy efficiency is less than that given in the simulations.

5.4.2 Fuzzy Equalization Based on State of Charge

Based on the equalizing circuit with the parameters given earlier, the equalizing process can be driven by the control

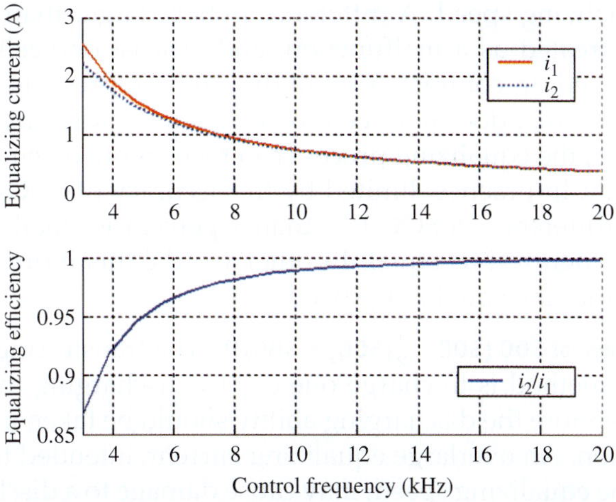

FIGURE 5.51 The average equalizing currents and equalizing efficiencies of various control frequencies ($SOC_1 = 0.6$, $SOC_2 = 0.5$). The energy efficiency only consider the energy loss on resistors.

frequency to MOSFET. Besides the speed of the equalizing process, the performance of an equalizer is also evaluated by the equalizing efficiency and the protection of cells. Therefore, a fixed control frequency is insufficient to achieve a satisfactory overall performance. Given the states of cells, how to tune the control frequency—and equally how to tune the equalizing current—becomes the key issue in equalizing system design.

5.4.2.1 Selection of Inputs

Generally, the selection of inputs of a controller should reflect the relevant factors, based on which the controller can determine an optimal, or at least a suitable, value of the control variable. The relevant factors affecting equalizing performance in this study are the equalizing speed, efficiency, and cell protection. Therefore, three corresponding cell states are utilized as the inputs.

The Difference in SOC ($\Delta SOC = |SOC_1 - SOC_2|$) The aim of battery equalization is for all cells to possess the same SOC. Cell imbalance is essentially the SOC imbalance rather than a voltage imbalance. A large ΔSOC indicates a serious situation and requires a comparatively large equalizing current to accelerate the equalizing speed. A difference in SOC larger than 30 percent is treated as a malfunction, and at least one cell has to be replaced. In contrast, a small difference represents a slight imbalance that does not need to be equalized urgently. In this situation, the equalizing process can focus more on equalizing efficiency. In practice, limited by the accuracy of SOC estimation, a difference in SOC less than 2 percent is small enough to be recognized as equal. In other words, this input mainly determines the equalizing speed.

The Average of SOC ($\overline{SOC} = \frac{1}{2}(SOC_1 + SOC_2)$) The transferring equalization method is to charge one cell by discharging the other one; therefore the discharging ability should be taken into consideration. An overlarge equalizing current, intended to accelerate the equalizing speed, may cause damage to a discharging cell with a low SOC. The average SOC represents the level of residual capacity. Therefore, it is used to indicate whether a large equalizing current is allowed. Usually, if the average SOC is less than 0.15, cells are in a very low state and a large equalizing current is strictly forbidden. Simply speaking, the second input chiefly focuses on the protection of cells.

The Total Internal Resistance ($\Sigma R = R_1 + R_2$) As shown in Fig. 5.47, the internal resistance varies according to the SOC. Meanwhile, as discussed previously, the resistance causes energy loss in the equalizing process. Therefore, a simple way of enhancing the equalizing efficiency is to apply a large current when the total internal resistance is small, and to apply a small current in the opposite situation. Thus, the third input focuses on energy efficiency.

5.4.2.2 Implementation of Fuzzy Controller

Based on the inputs proposed earlier, a controller should map the input vector to a control frequency, i.e., establish a function $F: f_s = F(\Delta SOC, \overline{SOC}, \Sigma R)$. However, because of the lack of quantitative analysis and modeling of the relationships among these variables, it is hard to design a controller based on traditional methods. In contrast, the inputs can easily be qualitatively classified into different levels and expressed as linguistic variables, such as large, small, and so on. There is, therefore, a strong rationale for applying fuzzy control to determine the equalizing current.

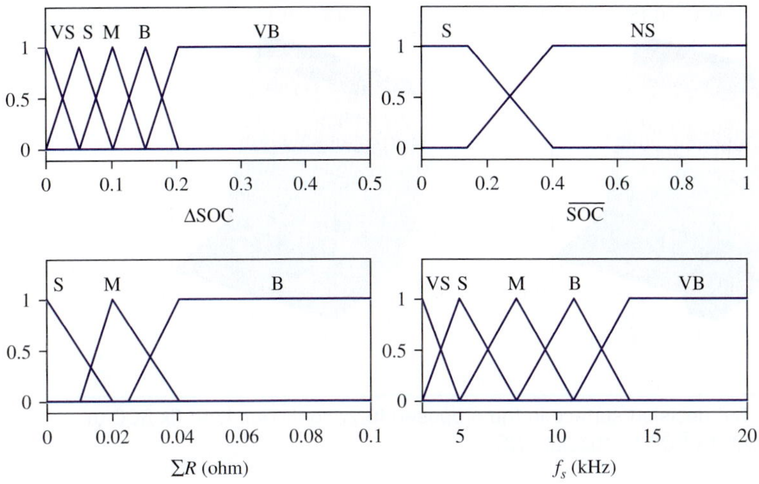

FIGURE 5.52 The membership functions of input and output variables in the proposed fuzzy controller.

In general, a fuzzy controller consists of the membership functions of inputs and outputs, a rule database, and inference methods.[80] The inference methods are the same as the default settings in the fuzzy logic toolbox given by MATLAB. As shown in Fig. 5.52, the membership functions of inputs and outputs are based on the experiences and basic knowledge of cell equalization, where VS = very small, S = small, M = middle, B = big, VB = very big, and NS = not small. The rule database is provided in Table 5.5, and hence the typical decision surfaces when $\overline{SOC} = 0.15$ and $\overline{SOC} = 0.4$ are shown in Fig. 5.53.

f_s	$\overline{SOC}$	S			NS		
	ΣR	S	M	B	S	M	B
ΔSOC	VS	VB	VB	VB	M	B	VB
	S	B	VB	VB	S	M	B
	M	M	B	VB	VS	S	M
	B	S	M	B	VS	VS	S
	VB	VS	S	M	VS	VS	VS

TABLE 5.5 Rules Database in the Proposed Fuzzy Controller

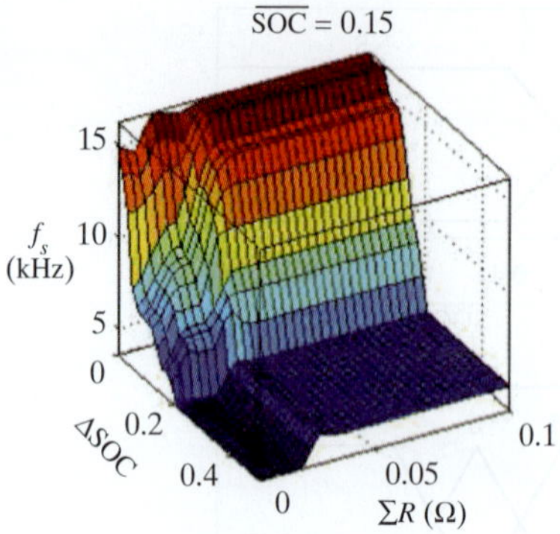

FIGURE 5.53 The decision surface in the proposed fuzzy controller. Left: fix $\overline{SOC}$ to absolute S; right: fix $\overline{SOC}$ to absolute NS.

5.4.3 Application Results

5.4.3.1 Simulation Results

To demonstrate the availability and efficacy of the proposed equalizing system, simulations are conducted using the 6-Ah battery model with features as shown in Figs. 5.46 and 5.47. The control period of the fuzzy controller is set to 1 s. During each second, the variation in cell states is small enough to be ignored so that only one MOSFET control cycle needs to be calculated, based on the given f_s, and the following $f_s - 1$ cycles will repeat the same dynamics. Thus, the total quantity of electricity transferred in 1 s equals f_s times that in any one cycle. Based on the accumulated variations, the SOC can be updated every second, and cell terminal voltages and internal resistances are changed according to their respective characteristic curves, as indexed by SOC. In the fuzzy controller, the inputs ΔSOC, $\overline{\text{SOC}}$, and ΣR are also calculated every second. Consequently, the MOSFET control frequency f_s will be tuned to realize an adaptive control and reach a satisfactory overall equalizing performance. The simulation temperature is fixed to 25°C. Considering the accuracy of SOC estimation in practice, the equalizing process will terminate when the SOC difference is smaller than 2 percent.

Four typical equalizing processes are illustrated in Fig. 5.54. The first two rows show the results under the initial conditions of $SOC_1 = 0.2$, $SOC_2 = 0.1$ and of $SOC_1 = 0.9$, $SOC_2 = 0.8$.

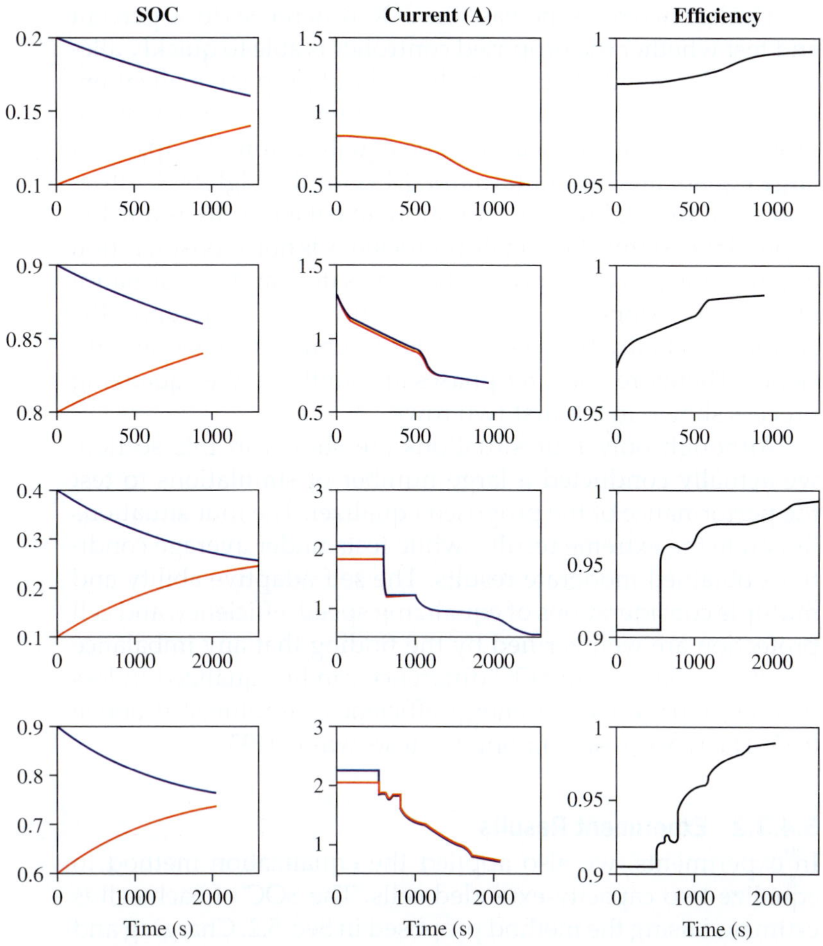

FIGURE 5.54 The simulation results under various initial conditions. Each row shows curves of SOC, equalizing current, and energy efficiency under the same initial conditions.

Both of them are initialized with a comparatively small SOC difference. Although the ΔSOC is the same in the beginning, the large total resistance and the limited discharging ability of low SOC cells restrict the equalizing current smaller than that in the situation of high SOC level. Therefore, the equalizing period in the low-SOC situation is longer, but with the benefit of high energy efficiency.

The last two rows increase the SOC differences to 30 percent and test whether the proposed controller is able to quickly alleviate the large imbalance, which, without proper equalization, risks damaging the cells by overcharging or overdischarging. From the curves, it is clear that the equalizer initially applies a large equalizing current (around 0.4 C for the high-SOC situation, 0.3 C for the low-SOC situation) to balance cells as quickly as possible. At this stage, energy efficiency is not a consideration because it is less of a priority than the safety and performance of the cells. Along with the reduction in SOC difference, the current gradually becomes smaller to emphasize energy efficiency. Therefore, the later phases are similar to the equalizing process shown in the first two rows.

Although only four situations are shown in this section, we actually conducted a large number of simulations to test the performance of the proposed equalizer. The four situations illustrate the extreme results, while tests under average conditions obtained moderate results. The self-adaptive ability and multiple considerations of equalizing speed, efficiency, and cell protection are well verified by the finding that any imbalance less than a 30 percent SOC difference can be equalized in less than 1 h, and the average energy efficiencies are almost all above 0.95. The worst result among the tests was 0.9493.

5.4.3.2 Experiment Results

In experiments, we also applied the equalization method to equalize two capacity-extended cells. The SOC of each cell is estimated using the method proposed in Sec. 5.2. Charging and discharging resistances are also modeled in Sec. 5.2.

Figures 5.55 and 5.56 show the equalizing cycles when the control frequencies of MOSFET are 5 and 20 kHz, respectively, obtained by oscillograph. Compared with the theoretical situation, the equalizing current and equalizing efficiency are both smaller. Figure 5.57 illustrates the average equalizing currents and equalizing efficiencies of various control frequencies in the experiments.

In the experiments, we applied three extended cells to test the performance under different initial conditions. Figures 5.58 and 5.59 show the equalizing processes of high and low initial SOC conditions, respectively. The SOC curve is estimated by

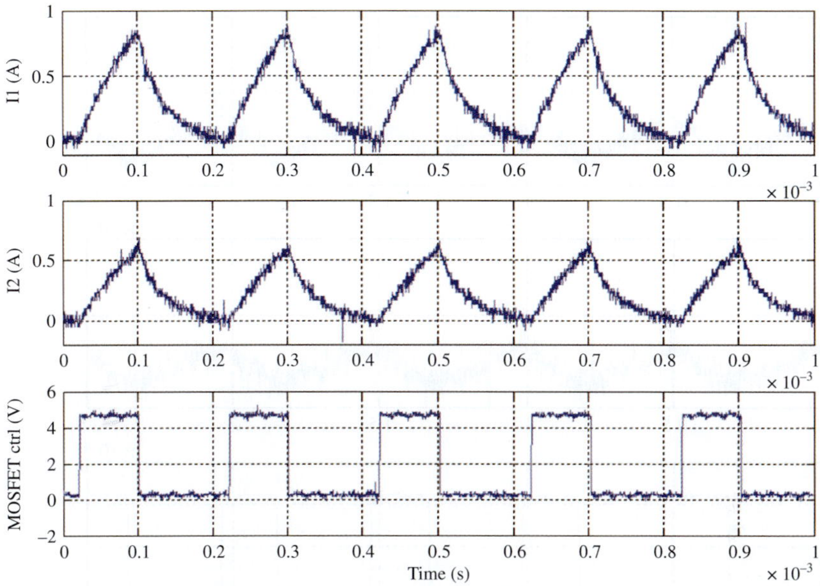

FIGURE 5.55 Equalizing cycles when control frequency of MOSFET is 5 kHz.

the method proposed in Sec. 5.2. Limited by the accuracy of SOC estimation in the experiment, we stop the equalization process when the largest SOC difference is below 5 percent. It is clear that, because the total internal resistance in the low-SOC condition is greater than that in the high-SOC condition, and the discharge ability of the cell with a low-SOC is worse than that of a high-SOC cell, the equalizing current is smaller in the low-SOC situation so as to generate less heat and protect cells.

The more significant performance is shown in Fig. 5.60, which illustrates the equalizing process in middle-SOC condition. Since $LiFePO_4$ has a very flat SOC versus OCV relationship, the large 20 percent SOC imbalance causes only around 20-mV difference in voltage. Along with the equalizing process, the voltage difference gradually becomes smaller than the error of the voltage measurement, which is around 8 mV in the experiment. Such a small voltage difference cannot be handled by the existing voltage-based fuzzy controllers. The method applies the SOC rather than the voltage as the input to the fuzzy

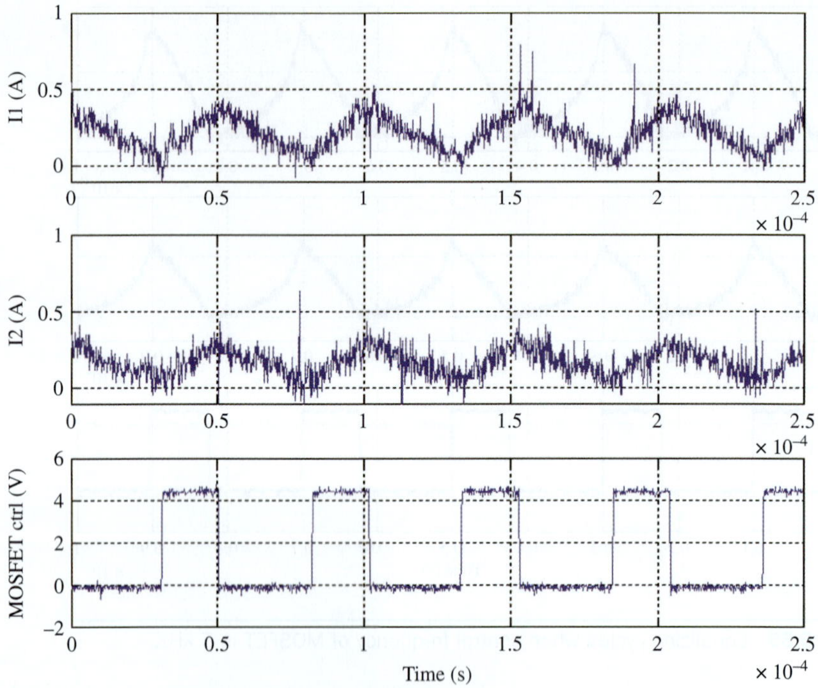

FIGURE 5.56 Equalizing cycles when control frequency of MOSFET is 20 kHz.

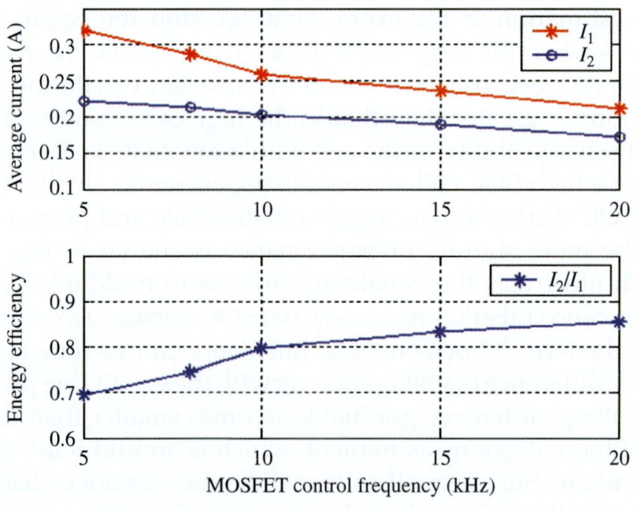

FIGURE 5.57 The average equalizing currents and equalizing efficiencies of various control frequencies in experiments.

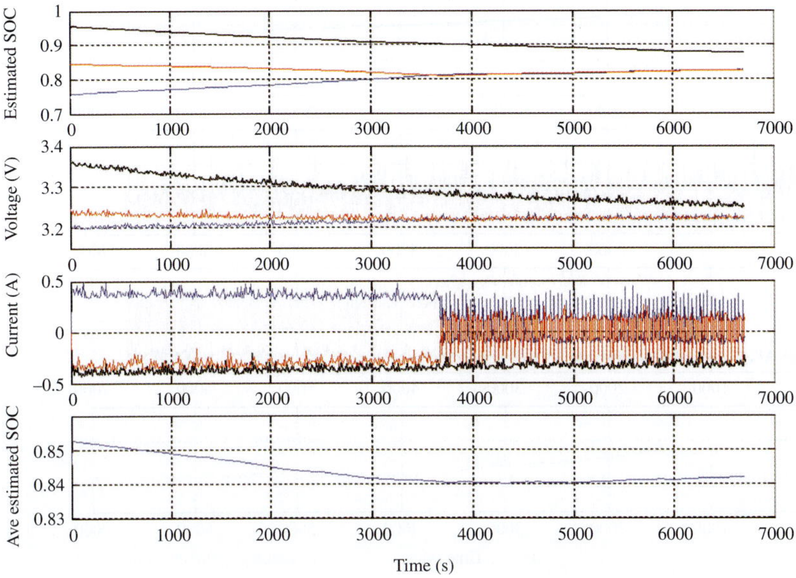

FIGURE 5.58 Equalization process in experiment with high-SOC initial condition (SOC = 0.95, 0.75,0.85).

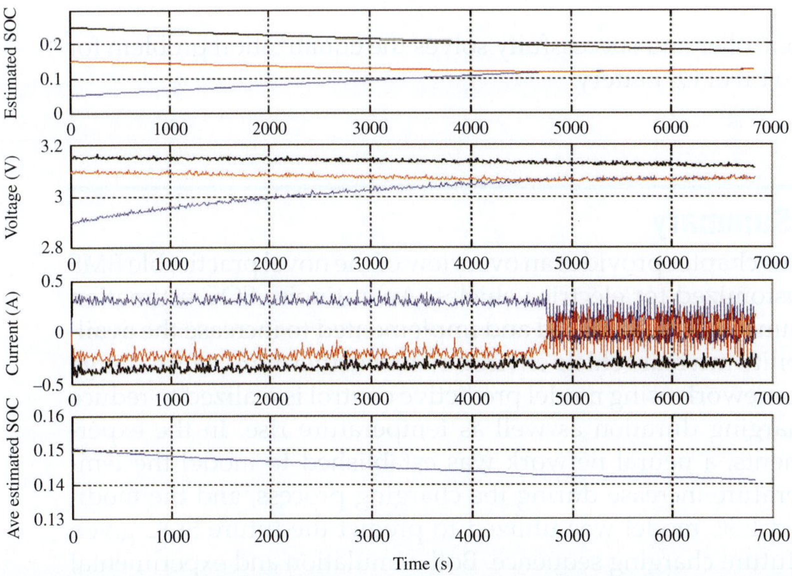

FIGURE 5.59 Equalization process in experiment with high-SOC initial condition (SOC = 0.25, 0.05,0.15).

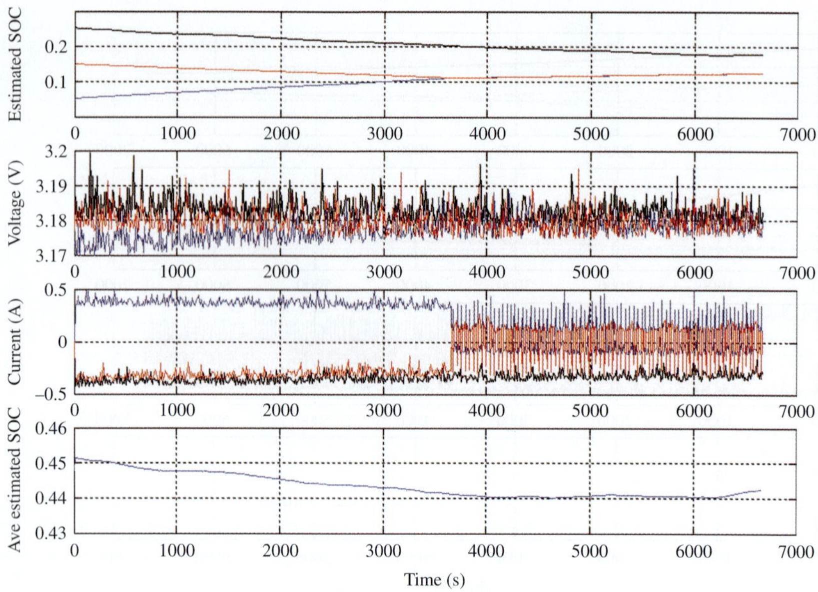

FIGURE 5.60 Equalization process in experiment with middle-SOC initial condition (SOC = 0.55, 0.35, 0.45).

controller and successfully solves the equalization problem for the $LiFePO_4$ battery.

5.5 Summary

This chapter provides an overview of the novel practicable BMS customized for electric vehicles. An antinoise SOC estimation framework is proposed and implemented to increase the availability in a real vehicle environment. Furthermore, a fast-charge framework using model predictive control is realized to reduce charging duration as well as temperature rise. In the experiments, a neural network was established to model the temperature increase during the charging process, and the modified ESC model was utilized to predict the future SOC given a future charging sequence. Both simulation and experimental results show that the MPC-based method can simultaneously reduce the charging duration and limit temperature increase.

The Pareto front of the proposed method has been demonstrated to dominate that of the most popular CCCV method. Finally, a SOC-based fuzzy controller is also proposed and implemented to equalize cells. Simulation and experimental results both show that the proposed controller has the ability to incorporate the experiences and knowledge of experts, as well as intelligently tune the charging current along with the instantaneous battery states to balance multiple objectives.

CHAPTER 6

Energy Management System

This chapter introduces the energy management system for HEV. Section 6.1 describes the model of the series electric vehicle. Section 6.2 presents the research of the load forecasting system and the simulation works. The cascade neural network with node-decoupled extended Kalman filtering (CNN-NDEKF) is applied to solve the load forecasting problem. Section 6.3 proposes the real-time power control system by combining the MPC-based power controller and the load forecasting system. Finally, the implementation and experiments of the proposed system are presented in Sec. 6.4.

6.1 Modeling of Series Electric Vehicle

By definition, the HEV is a vehicle in which propulsion energies are available from two or more kinds or types of energy stores, sources, or converters, and at least one of them can deliver electric energy. Based on this general definition, there are many configurations of HEVs, such as the gasoline ICE and battery, diesel ICE and battery, battery and capacitor, battery and flywheel, and battery and battery hybrids. In this book, the term *HEV* refers only to the vehicle utilizing the gasoline ICE, electric motor, and battery.

The *traction system* in a hybrid electric vehicle can be defined as the components generating or transforming power needed for the powertrain. Depending on the ways in which the components are chosen and arranged, hybrid electric vehicles can be classified into three topologies. Primitively, HEVs are classified into two basic topologies—serial and parallel. The main difference between the two powertrain configurations is the connection between the power components and the wheels. Recently, a new series-parallel hybrid vehicle, which integrates the features of both the series and parallel hybrids, has been added to the list of HEV topologies. In the series hybrid, the ICE is connected with the generator to produce electricity, and no mechanical connection is made between the ICE and the wheels, resulting in a pure-electric propulsion. In the parallel hybrid, both the ICE and the electric motor (EM) are connected with the wheels to propel the vehicle. In the series-parallel hybrid, both series and parallel ways can be used to propel the vehicle for different driving conditions.

Although there are three kinds of HEV configuration, this book focuses on only the series hybrid system. In a series hybrid system, as shown in Fig. 6.1, the mechanical power output from

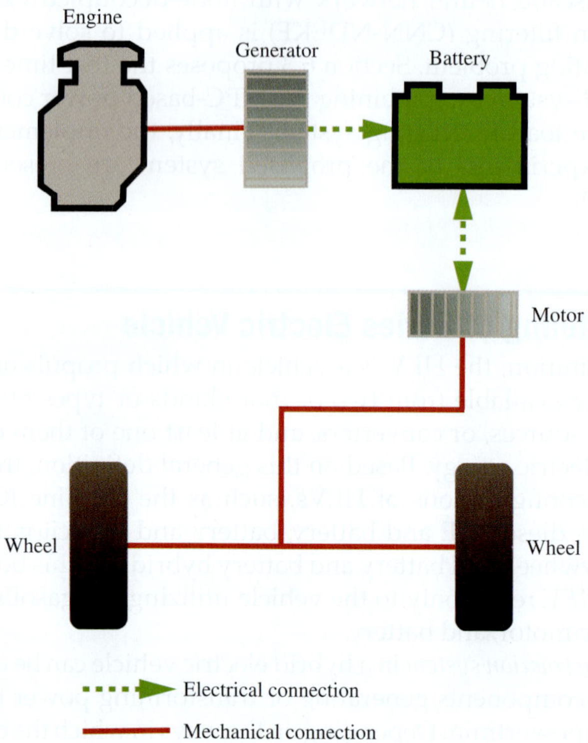

FIGURE 6.1 Series HEV configuration.

the ICE is converted to electric energy as the generator. The electricity is either used to charge the battery or bypassed from the battery to the motor, which propels the wheels.

The advantage of the series configuration is that the ICE is running mostly at its optimal point of speed and torque, thereby having a low fuel consumption and high efficiency. However, there are two energy conversion stages during the transformation of the energy between the ICE and the wheels (ICE-generator and generator-motor). Some energy is lost to the two-stage power conversion process. A series hybrid vehicle is more applicable for city driving.

A hybrid vehicle consists of some complex components. The dynamics of the different components in the system range from *s* in the power electronics to months for the aging of batteries. This is not an acceptable time span to encapsulate in one model.

The purpose of the modeling is to achieve a model of a hybrid vehicle that is well suited for calculation of power flows. Of course accuracy should be as good as possible, but an error in the fuel consumption or battery loss of a few percent is tolerable as long as this behavior is consistent. When the focus is on power flows in an HEV, it is not interesting to look at dynamics that are much faster than the system-level dynamics of the HEV; thus the dynamics that are much faster than 1 Hz could be ignored.

6.1.1 Main Components

The general model of an HEV is based on key components or subsystems such as the engine, transmission, motor/generator, battery, wheel, and vehicle dynamics. In this section, all these components are described, and the series hybrid 4WIS is constructed.

6.1.1.1 Engine

The engine is modeled with a static model. The fuel consumption is a static map of two independent variables: engine speed and engine torque. We use static maps for engine modeling for these reasons:

1. The dynamics during one combustion-expansion cycle are fast (milliseconds) and thus not interesting in this power management case.

	First Engine	Second Engine
Model	CF172MM	BENZ A160
Type	Single-cylinder, 4-stroke	Four-cylinder, 4-stroke
Engine Capacity	250 cc	1600 cc
Displacement	244 mL	1598 mL
Max. power	11 kW/4500 rpm	150 kW/4000 rpm
Max. torque	18 N · m/4500 rpm	75 N · m/5250 rpm
Ignition	C.D.I	S.I

TABLE 6.1 Engine Specifications

2. When the operating points of the ICE are changed slowly (seconds), fuel consumption from static maps is accurate enough.

In this chapter, the series HEV are built up based on two different engines. Table 6.1 shows some main specifications. The engines are described with static maps that give fuel efficiency as functions of speed and torque. In this chapter, the engines are assumed to be fully warmed up; thus, the engine temperature effect is not considered.

Figures 6.2 and 6.3 depict the efficiency maps of the engines.

Figure 6.4 shows the maximum power and torque curve of the second engine. The curves are used to confirm the maximum power that the engine can supply.

6.1.1.2 Motor/Generator

The electric motors considered for hybrid electric vehicles are either dc motors or brushless dc motors.[8, 47] In this chapter, brushless dc (BLDC) motors are chosen which can be used motors and generators.

The model is based around the trapezoidal back electromotive forces (emf's), star wound stator-type brushless dc motor. The modeling of the brushless dc motor involves solving some simultaneous differential equations. We treat some parameters as constant, such as phase inductance, torque constants, and inner resistance. The core for the brushless dc motor has been written as a state space model.

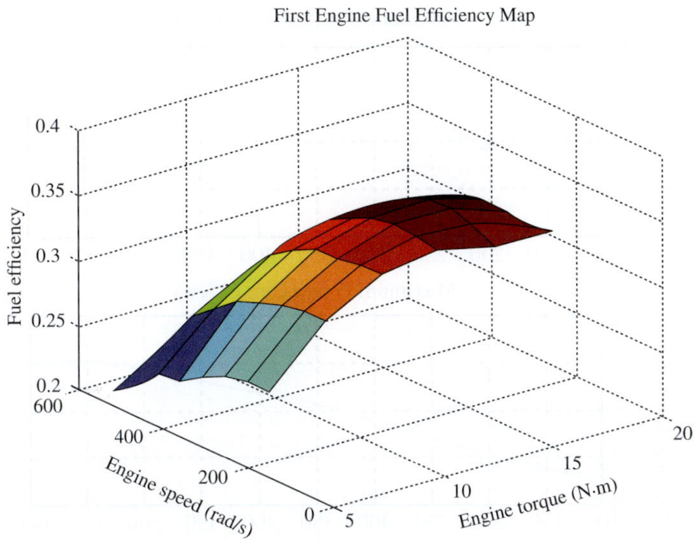

FIGURE 6.2 First engine fuel efficiency map.

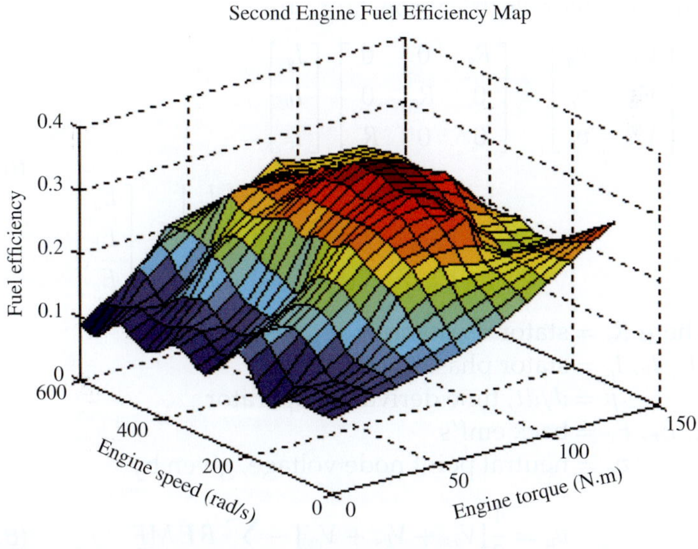

FIGURE 6.3 Second engine fuel efficiency map.

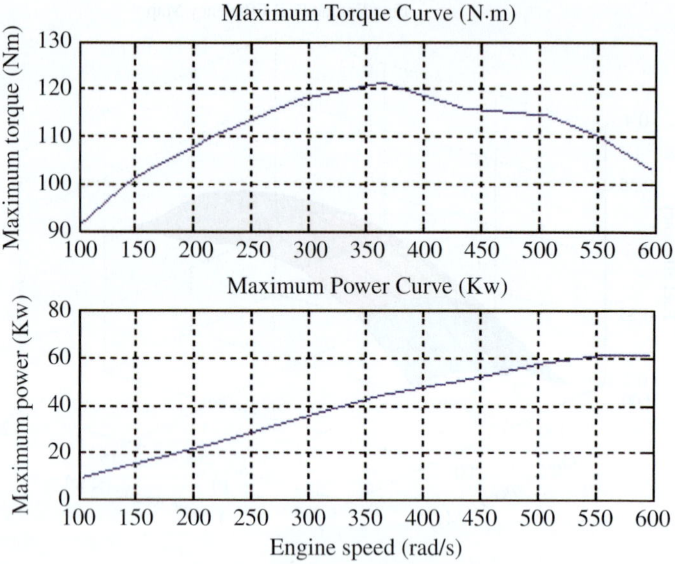

FIGURE 6.4 Second engine maximum torque and power curve.

The coupled circuit equations of the stator windings in terms of motor electrical constants are

$$\begin{bmatrix} V_{as} - v_n \\ V_{bs} - v_n \\ V_{cs} - v_n \end{bmatrix} = \begin{bmatrix} R_s & 0 & 0 \\ 0 & R_s & 0 \\ 0 & 0 & R_s \end{bmatrix} \begin{bmatrix} I_a \\ I_b \\ I_c \end{bmatrix} + p \begin{bmatrix} L_{aa} & L_{ab} & L_{ac} \\ L_{ba} & L_{bb} & L_{bc} \\ L_{ca} & L_{cb} & L_{cc} \end{bmatrix} \begin{bmatrix} I_a \\ I_b \\ I_c \end{bmatrix} + \begin{bmatrix} E_a \\ E_b \\ E_c \end{bmatrix} \tag{6.1}$$

where R_s = stator resistance per phase
I_a, I_b, I_c = stator phase currents
$p = d/dt$, time derivative operator
E_a, E_b, E_c = back emf's
v_n = neutral point node voltage, given by

$$v_n = \frac{1}{3}[V_{as} + V_{bs} + V_{cs}] - \sum BEMF \tag{6.2}$$

Here $\sum BEMF$ means the summation of the individual phase emf's on an instant-to-instant basis. The induced emf's are all assumed to be trapezoidal, whose peak value is given by

$$E_p = (BLv)N = N(BLr\omega) = N\phi\omega = \lambda\omega \tag{6.3}$$

where B = flux density of the field, in webers; L = rotor length; N = number of turns per phase; ω = electrical angular speed in rad/s; ϕ = flux linkage = BLr; λ = total flux linkage, given as the product of the number of conductors and flux linkage per conductor.

We assume that (1) there is no change in rotor reluctance with angle because of the nonsalient rotor and three symmetric phases and (2) inductances and mutual inductances are symmetric for all phases. Then Eq. (6.1) becomes

$$\begin{bmatrix} V_a \\ V_b \\ V_c \end{bmatrix} = \begin{bmatrix} R_s & 0 & 0 \\ 0 & R_s & 0 \\ 0 & 0 & R_s \end{bmatrix} \begin{bmatrix} I_a \\ I_b \\ I_c \end{bmatrix} + p \begin{bmatrix} L & M & M \\ M & L & M \\ M & M & L \end{bmatrix} \begin{bmatrix} I_a \\ I_b \\ I_c \end{bmatrix} + \begin{bmatrix} E_a \\ E_b \\ E_c \end{bmatrix} \tag{6.4}$$

Equation (6.4) can be further simplified to

$$\begin{bmatrix} V_{as} \\ V_{bs} \\ V_{cs} \end{bmatrix} = \begin{bmatrix} R_s & 0 & 0 \\ 0 & R_s & 0 \\ 0 & 0 & R_s \end{bmatrix} \begin{bmatrix} I_a \\ I_b \\ I_c \end{bmatrix} + p \begin{bmatrix} L-M & 0 & 0 \\ 0 & L-M & 0 \\ 0 & 0 & L-M \end{bmatrix} \times \begin{bmatrix} I_a \\ I_b \\ I_c \end{bmatrix} + \begin{bmatrix} E_a \\ E_b \\ E_c \end{bmatrix} \tag{6.5}$$

The generated torque is given by

$$T_e = [E_a I_a + E_b I_b + E_c I_c]/\omega \tag{6.6}$$

The back emf's—E_a, E_b, E_c—are voltages proportional to the angular rate ω seen at the shaft

$$\begin{aligned} E_a &= f_a(\theta)\lambda\omega \\ E_b &= f_b(\theta)\lambda\omega \\ E_c &= f_c(\theta)\lambda\omega \end{aligned} \tag{6.7}$$

where, $f_a(\theta)$, $f_b(\theta)$, and $f_c(\theta)$ are functions that have same shapes as back emf's with magnitude from −1 to 1.

The mechanical part of the motor equations is derived using Newton's law, which states that the inertial load J times the derivative of angular rate equals the sum of all the torques about the motor shaft. The result is

$$J\frac{d\omega}{dt} = \sum \tau_i = -K_f\omega + T_e - T_l \tag{6.8}$$

where T_l = load torque, J = moment of inertia, and K_f = friction coefficient.

Combining all the equations, the BLDC system can be described as a state space form

$$\begin{aligned} x' &= Ax + Bu \\ y &= Cx + Du \end{aligned} \tag{6.9}$$

where

$$x = \begin{bmatrix} I_a & I_b & I_c & \omega \end{bmatrix}^T \tag{6.10}$$

$$u = \begin{bmatrix} V_a & V_b & V_c & T_l \end{bmatrix}^T \tag{6.11}$$

$$y = \omega \tag{6.12}$$

The state space matrices are

$$A = \begin{bmatrix} -R_s/L_1 & 0 & 0 & \lambda f_a(\theta)/L_1 \\ 0 & -R_s/L_1 & 0 & \lambda f_b(\theta)/L_1 \\ 0 & 0 & -R_s/L_1 & \lambda f_c(\theta)/L_1 \\ \lambda f_a(\theta)/J_1 & \lambda f_b(\theta)/J_1 & \lambda f_c(\theta)/J_1 & -K_f/J_1 \end{bmatrix} \tag{6.13}$$

$$B = \begin{bmatrix} 1/L_1 & 0 & 0 & 0 \\ 0 & 1/L_1 & 0 & 0 \\ 0 & 0 & 1/L_1 & 0 \\ 0 & 0 & 0 & -1/J_1 \end{bmatrix} \tag{6.14}$$

$$C = \begin{bmatrix} 0 & 0 & 0 & 1 \end{bmatrix} \tag{6.15}$$

and

$$D = \begin{bmatrix} 0 & 0 & 0 & 0 \end{bmatrix} \tag{6.16}$$

where $L_1 = L - M$, L = self-inductance per phase, M = mutual inductance per phase, and V_a, V_b, V_c = impressed voltages on the motor windings per phase. The BLDC model presented above can operate in either generator or motor mode. The mode of operation is dictated by the sign of the load mechanical torque T_l (positive for motor mode, negative for generator mode).

6.1.1.3 Battery

As discussed in Chap. 5, batteries are probably the most difficult to understand and model of all the components in a hybrid electric vehicle. Although acting as simple electric energy storage devices, when they deliver and accept energy, batteries actually undergo thermally dependent electrochemical processes that make them difficult to model. The electrical behavior of a battery is a nonlinear function of a variety of constantly changing parameters such as voltage, current, temperature, and efficiency. To model batteries is to find a compromise between trying to include all the relevant effects and creating a model that will actually work in a reasonable range of charging parameters.

Figure 6.5 shows the charging character of the battery. It can be seen that charging efficiency decreases when SOC is high. Thus, batteries would better be charged at low SOC levels. Similarly, the batteries discharge is inefficient when SOC is low. Therefore, a predefined operational range is necessary for batteries.

In this chapter, a *RC* model similar to that in Chap. 5 is used to develop the battery model. The *RC* model includes transient effects in the batteries. The *RC* model uses the charging/discharging current as the input and the battery voltage as

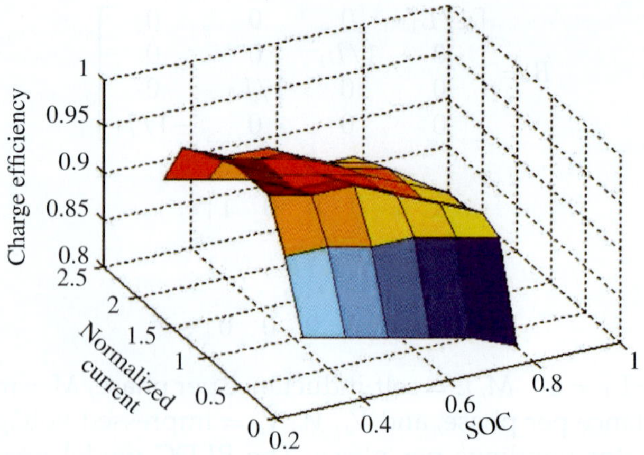

FIGURE 6.5 Battery charge efficiency with SOC and current at 25°C.

the output which is limited to the operational range. In addition, the ambiance temperature is assumed to be 25°C and constant.

The *RC* model is shown in circuit form in Fig. 6.6. It consists of a bulk capacitor C_b and a surface capacitor C_s, which simulate energy storage in the battery and the dynamic properties of the battery, respectively. Output resistance R_o, surface resistance R_s, and bulk resistance R_b are used to model the internal

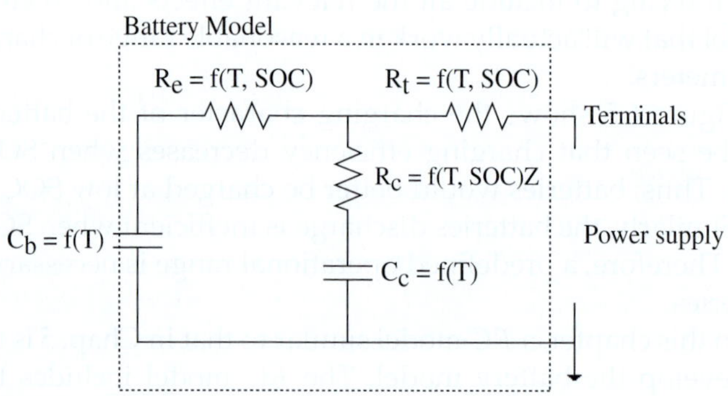

FIGURE 6.6 *RC* battery model.

resistance of battery. The governing state equations of the model are as follows:

$$\begin{bmatrix} \dot{V}_{cb} \\ \dot{V}_{cc} \end{bmatrix} = \begin{bmatrix} -1/C_b(R_e + R_c) & 1/C_b(R_e + R_c) \\ 1/C_c(R_e + R_c) & -1/C_c(R_e + R_c) \end{bmatrix} \begin{bmatrix} V_{cb} \\ V_{cc} \end{bmatrix} + \begin{bmatrix} -R_c/C_b(R_e + R_c) \\ -R_e/C_b(R_e + R_c) \end{bmatrix} [I_s] \tag{6.17}$$

$$[V_o] = \begin{bmatrix} -R_c/(R_e + R_c) & -R_e/(R_e + R_c) \end{bmatrix} \begin{bmatrix} V_{cb} \\ V_{cc} \end{bmatrix} + [-R_t - R_c R_e/(R_e + R_c)][I_s]$$

SOC Estimation SOC estimation is one of the most important issues to the applications of the battery. Precise SOC estimation allows an efficient utilization, and long-life use of the battery.

By monitoring the voltage, charging/discharging current, and temperature of the batteries, we use a Kalman filter to estimate the state of the battery, that is, V_{cb} and V_{cc}. Then the SOC of the battery is obtained from

$$\text{SOC} = \frac{20V_{cb}C_b + V_{cc}C_c}{21Q_f} \tag{6.18}$$

where Q_f is the battery's rated power. The weights for the bulk capacitor C_b and surface capacitor C_s contributions to overall SOC are 20 and 1.

6.1.1.4 Automatic Transmission

The automatic transmission transmits torque from the engine to the drive axle. It is composed of a hydraulic torque converter and a gearbox.

Torque Converter A torque converter couples two axes and transfers torque and angular motion by the hydrodynamic action of a viscous fluid. Unlike a friction clutch, it cannot lock the axes together. It is usually installed between the engine and the gearbox in vehicles equipped with conventional automatic transmissions. It can increase the engine's output torque, decrease the output speed, and dampen vibrations.

Three functions specify the characteristics of the torque converter.[13]

$$T_p = \left[\frac{\omega_e}{K(\omega_r)}\right]^2, \qquad T_t = T_r(\omega_r)\cdot T_p \tag{6.19}$$

$$T_x = [T_t - T_{x,l}(\omega_t, g_x)]\cdot R_x(g_x)\cdot \eta_x(g_x) \tag{6.20}$$

$$T_d = [T_x + R_c\cdot T_m\cdot \eta_c - T_{d,l}(\omega_x)]\cdot R_d\cdot \eta_d \tag{6.21}$$

where T_p and T_t are the pump and turbine torque; K and T_r are the capacity factor and torque ratio of the torque converter; $\omega_r = \omega_t/\omega_e$ is the speed ratio of the torque converter; T_x and T_d are the output torques of the transmission and differential, respectively; $T(x, l)$ and $T(d, l)$ are the torque loss due to friction and churning loss for the transmissions and differential; R_x and η_x are the gear ratio and efficiency of the transmission, respectively, which are functions of the gear number g_x.

Similarly with the engine, because the torque converter response is fast, static curves are used to model K and T_r, as shown in Fig. 6.7.

Planetary Gear Sets and Gearbox Shift Control A Simpson planetary gear set (PGS) is used after the torque converter. The PGS can realize four different forward gear ratios listed in Table 6.2.

The main function of the gear shift control is to use the throttle angle and the transmission output speed to determine when to shift up or down. A number of shift patterns are used in automatic transmission. To make the power loss as low as possible, a shift logic used in this chapter is shown in Fig. 6.8.[118]

6.1.1.5 Wheel

The wheel is a flexible body contacting the road surface and subjected to slip.[28] When an input torque is applied to the wheel axle, the tire deforms, pushes on the ground (while subject to contact friction), and transfers the resulting reaction as a force back on the wheel, pushing the wheel forward or backward. Because this chapter focuses on the power control of the HEV, the wheel model includes only longitudinal motion and no camber, turning, or lateral motion.

A wheel model must provide the longitudinal force F_x the tire exerts on the wheel given the vertical load F_z and contact

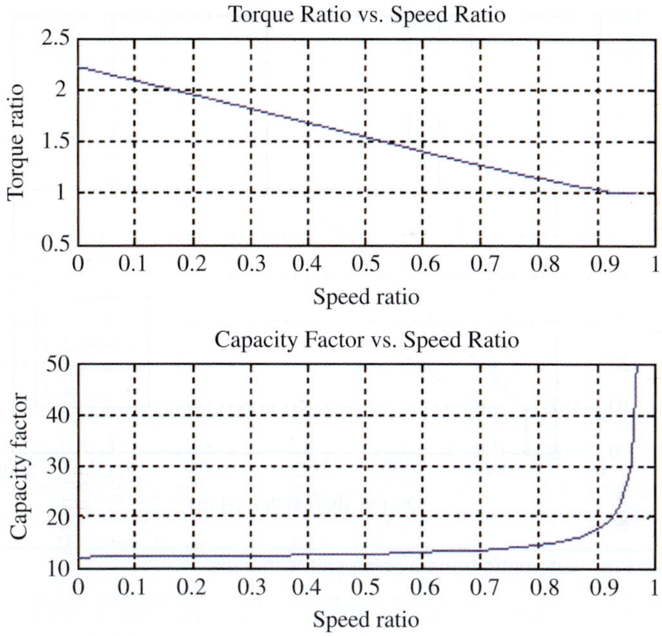

FIGURE 6.7 Torque ratio and capacity factor on different speed ratios of the torque converter.

slip k'. At full speed, the tire acts as a damper, and the longitudinal force F_x is determined mainly by the slip. At low speeds, when the tire is starting up from or slowing down to a stop, the tire behaves more as a deformable, circular spring. The effective rolling radius r_e is normally slightly less than the nominal tire radius because the tire deforms under its vertical load. The tire relaxation length σ_k is the ratio of the slip stiffness to

Gear	Gear Ratio
1st	2.4
2nd	1.45
3rd	1
4th	0.875

TABLE 6.2 Gear Ratios

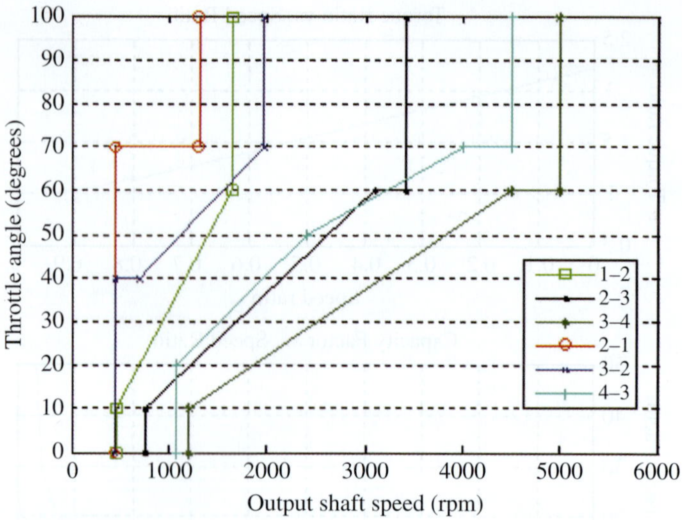

FIGURE 6.8 Automatic transmission shift schedule.

longitudinal force stiffness. It determines the transient response of F_x to slip. A static characteristic function specifies this relationship: $F_x = f(k', F_z)$.

6.1.1.6 Vehicle

The classic equation for longitudinal vehicle dynamics [28,76] is

$$
\begin{aligned}
m\dot{V}_x &= F_x - F_d - mg\sin\beta \\
F_x &= F_{xf} + F_{xr} \\
F_d &= \frac{1}{2}C_d\rho A V_x^2 \cdot \text{sgn}(V_x)
\end{aligned}
\tag{6.22}
$$

where F_x is the force from the wheels to drive the vehicle, F_d is aerodynamic drag, and $mg\sin\beta$ is the force of gravity that must be overcome to climb a grade. Here β is the incline angle (rad), m is vehicle mass (kg), A is effective frontal vehicle cross-sectional area (m^2), and V_x is the longitudinal vehicle velocity (m/s); C_d is the aerodynamic drag coefficient ($N \cdot s^2/kg \cdot m$) and $\rho = 1.2\,kg/m^3$ is the mass density of air.

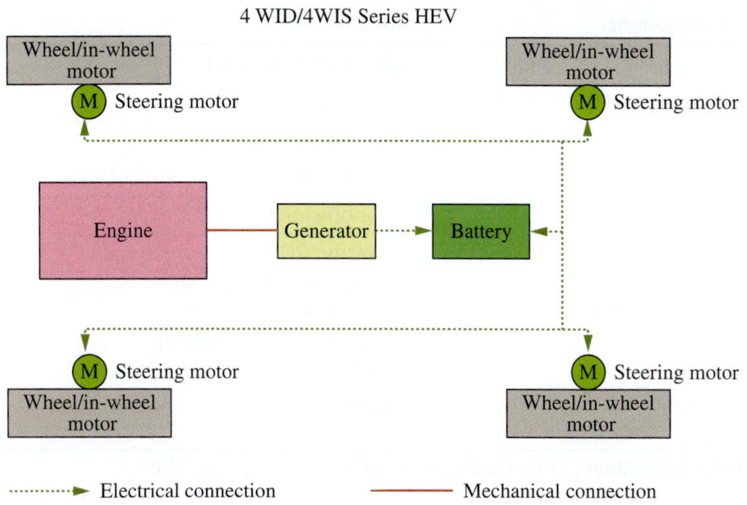

FIGURE 6.9 The 4WIS series hybrid electric vehicle structure.

6.1.2 4WIS Series Electric Vehicle

The 4WIS series electric vehicle is built up as in Fig. 6.9. It is a series HEV with an engine/generator system to supply electric power stored in a battery package. Four in-wheel motors are used to drive the vehicle forward or backward. In addition, four steering motors are installed so that all four wheels can independently steer. A set of 20-Ah Ni-Mh batteries are used as an energy storage device.

Table 6.3 gives some main specifications of the engine, generator, in-wheel motor, and battery. Table 6.4 illustrates the drivability, operation range, etc.

6.1.3 Building Up and Analysis

In the earlier sections, the series HEV models are nonlinear because of the nonlinearity of main components. To apply the model predictive control method to the HEV power control problem, we need to linearize the HEV models and eliminate unobservable states in state-space models. Furthermore, we also need HEV models to simulate and validate the feasibility of

Component	Specifications
Engine	250 cc, C.D.I engine,
PM in-wheel motor (×4)	Max power: 1.2 kW
	Peak efficiency: 89%
PM generator	Max power: 3.5 kW
	Peak efficiency: 90%
Ni MH battery	Capacity: 20 Ah Number of modules: 4 Nominal voltage: 12 V Power density: 200 W/kg Energy density: 24 W·h/kg
Vehicle	Weight: 420 kg

TABLE 6.3 Main Component Specifications of the Series HEV

the proposed method. In this section, we build up HEV models and analyze them.

The series HEV model is built up by MATLAB/Simulink as shown in Fig. 6.10.

The model consists of three subsystems — engine-generator subsystem, wheel-vehicle subsystem, and battery subsystem. The inputs and outputs of the model are listed in Tables 6.5 and 6.6.

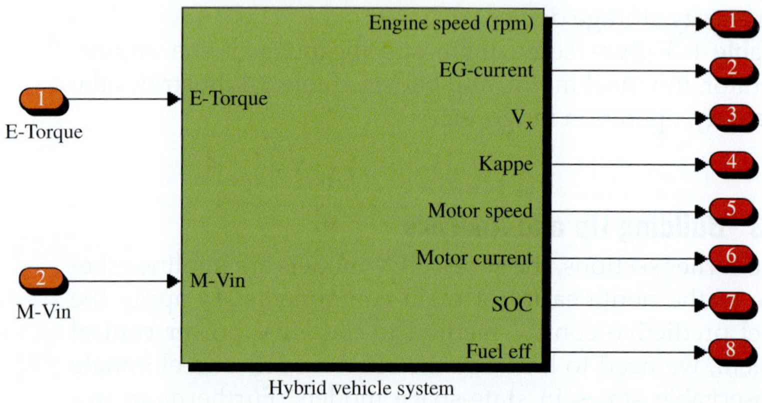

FIGURE 6.10 HEV model.

Specifications		Data
Drivability	Max speed (km/h) Acceleration time (0–40 km/h) (s) Gradeability (%)	$\geq$ 40 $\leq$ 18 15
Economy	Fuel consumption (kWh/20 km) Driving range (km)	$\leq$ 20 $\geq$ 40
Noise	Noise during acceleration (dB)	Less than standard 2–3
Reliability	Range (km)	$\geq$ 500

TABLE 6.4 Drivabilities of the Series HEV

6.1.3.1 States Definition

Identifying states of the model is important for analyzing a driveline, particularly a complex vehicle with many constraints and external inputs. In a vehicle model, the mechanical motions are all rotational. By searching every single, distinct angular velocity, we find the states list as in Table 6.7.

6.1.3.2 Linearizing Model

Model linearization needs to be done because the MPC is good at handling the nonlinear case. A MATLAB function *linmod* is used to linearize the model based on the predefined

No.	Name	Specification
1	*E-Torque*	Engine output torque to generator
2	*M-Vin*	Driving motor torque

TABLE 6.5 Input List of HEV Model

No.	Name	Specification
1	*Engine Speed*	Engine output shaft speed
2	*EG-Current*	Engine-generator system output current
3	V_x	Vehicle speed
4	*Kappa*	Kappa parameter of wheel
5	*Motor Speed*	Driving motor speed
6	*Motor Current*	Driving motor current
7	SOC	Battery state of charge
8	*Fuel eff*	Fuel efficiency

TABLE 6.6 Output List of HEV model

No.	States
1	Engine speed
2	Generator output current
3	Vehicle speed
4	Tire force derivative with respect to vertical load Fz
5	Driving motor speed
6	Driving motor current
7	Battery SOC
8	Fuel efficiency
9	Horizontal load force
10	Derivation of horizontal load force

TABLE 6.7 States List of Series HEV Model

states and input vector. During the control process, the linearization is being produced periodically based on real-time states and the input vector in order to enhance the control performance.*

6.1.3.3 Observability and Controllability of HEV Models

The model is a two-input-eight-output system. Obviously, the vehicle consists of some components that will affect the independence of states. For example:

1. Driving motor speed ($Motor\ Speed$) is always correlative with vehicle speed (V_x) for a series HEV.
2. A clutch, if unlocked, also connects two driveline axes and can impose a relative torque between them, leaving the two axes independent. If a clutch locks, applying only static friction between the two connected axes, the two axes are no longer independent.
3. When you engage the Simpson four-speed automatic transmission, two of its clutches lock, removing its 2 independent degrees of freedom (DoFs) and reducing the total from 4 to 2. They are the engine-to-converter axis

*The algorithm used for linearization can be found on the MathWorks web site at http://www.mathworks.com/access/helpdesk/help/toolbox/slcontrol/.

and the converter-to-road axis. There are still four driveline states, but two become dependent. When the transmission is engaged and the lockup clutch in the torque converter is also locked, only one independent DoF remains. This DoF is the entire drivetrain from engine to road.

To eliminate the unobservable state in the state-space model, a MATLAB function *minreal* is used to guarantee the observability of models.

The outputs of the system are greater than the inputs. Thus, it is impossible to hold all outputs at set points. In this case, we have two options:

1. Enforce set points on all outputs, in which case all will deviate from their set points to some extent.
2. Specify that certain outputs need not be held at set points by setting their weights to zero on the MPC controller's weight.

Outputs that have been sacrificed by setting their weights to zero can still be useful. If measured, they can help the controller to estimate the plant's state, thereby improving its predictions. They can also be used as indicators, or as variables to be held within an operating region defined by output constraints.

6.2 Load Forecasting

This section presents an approach for the driving load forecasting problem based on CNN-NDEKF. Because of the inherent advantages of increased fuel economy and better vehicle performance, HEV powered by ICE and energy storage are being given greater attention. The power control strategy of HEVs is the key technology that makes the HEVs more efficient. Since the extent of improvement included with HEV power control strategies greatly depends on the driving load forecasting, CNN-NDEKF is used to forecast the driving load in this section. Instead of forecasting the entire load sequence, we define five load levels by a fuzzy logic method and forecast what the load level will be in the future. A simulation study is given to illustrate the feasibility of the proposed forecasting approach.

In general, existing HEV power control approaches have the limitation that they used only the current vehicle state for decision making in power distribution; little consideration is generally given to the future load. Furthermore, the different driving style of the drivers and the trend of the vehicle will also affect the vehicle performance very much. The existing HEV power control approaches do not consider them either. The proposed power control approach incorporates a driving load forecasting algorithm whose role is to assess the driving environment, the driving style of the driver, and the trend of the vehicle using long- and short-term statistical features of the past drive cycle. After studying the methods for HEV power control strategy, we use CNN to forecast the future load of HEVs. The proposed driving load forecasting algorithm is to assess the driving environment, the driving style of the driver, and the trend of the vehicle using long- and short-term statistical features of the past drive cycle.

CNN-NDEKF does not rely on the rigid, fixed architecture networks and/or slow, gradient descent-based training algorithms because it is the composite of (1) flexible cascade neural networks (CNNs), which dynamically adjust the size of the neural network as part of the learning process, and (2) node-decoupled extended Kalman filtering (NDEKF), a fast converging alternative to back propagation. In this section, CNN-NDEKF is used to forecast the driving load because of its advantage of dealing with the nonlinear parts of the forecasted load and its fast converging speed.

6.2.1 Definition of Load Level

The driving load in a future time slot is proportional to the vehicle net wheel torque as calculated by

$$
\begin{aligned}
T_{T_s} &= \frac{1}{T_s}\int_0^{T_s} T_{wh(t)}\,dt \\
T_{wh}(t) &= a(t)M_r r_d + B_{wh}V(t) + \frac{V(t)r_d^2}{|V(t)|}[F_r + F_a(V(t))]
\end{aligned}
\tag{6.23}
$$

where T_{T_s} is the average torque of the future time slot T_s; $T_{wh(t)}$ is the net wheel torque at time t; $a(t)$ is the acceleration of the vehicle at time t; $M_r = M_v + J_r/r_d^2$ is the effective mass of

Load Level	Fuzzy Data
Very low	1
Low	2
Medium	3
High	4
Very high	5

TABLE 6.8 Fuzzy Set of Load Level Data

the vehicle; M_v is the mass of the vehicle; J_r is the equivalent moment of inertia of the rotating components in the vehicle; r_d is the dynamic tire radius, B_{wh} is the viscous damping, F_r and F_a are the rolling resistance force and the aerodynamic drag force, respectively. Obviously, the relationship between the driving load and the acceleration and speed of the vehicle is complex and nonlinear.

There are infinite kinds of future load sequences. Theoretically speaking, it might be better if we could forecast an entire load sequence; however, forecasting the entire load sequence will cost enormous computation and is not necessary. After considering the tradeoffs between computational cost and the HEV power control strategy requirement, in this section, we divide the driving load into five levels as shown in Table 6.8, and a fuzzy logic approach is used to determine the driving load level.

The average acceleration and speed are chosen as the input parameters of the fuzzy logic approach. They are divided into five levels, as shown in Tables 6.9 and 6.10.

Membership functions of the fuzzy sets are shown in Figs. 6.11 and 6.12. The fuzzy rules are represented in a linguistic

Average Speed Level	Range (mph)
Very low	0–20
Low	10–30
Medium	20–50
High	40–60
Very high	≥ 50

TABLE 6.9 Fuzzy Set of Average Speed Data

Average Acceleration Level	Range (ft/s^2)
Negative big	< −0.06
Negative small	−0.12–0
Zero	−0.06–0.06
Positive small	0–0.12
Positive big	> 0.06

TABLE 6.10 Fuzzy Set of Average Acceleration Data

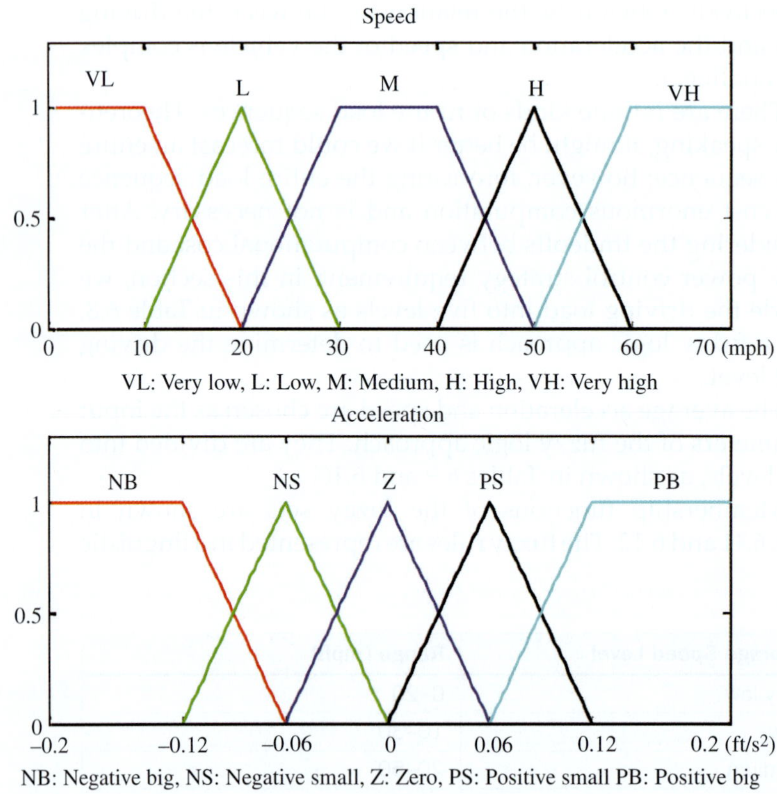

FIGURE 6.11 Membership function for input variables.

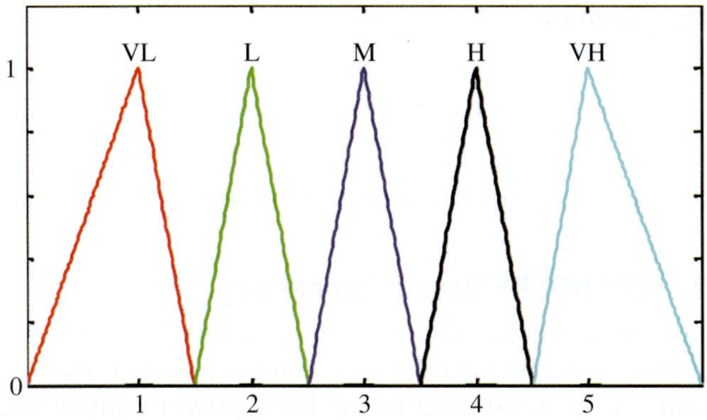

FIGURE 6.12 Membership function for output variables.

form in Table 6.11. The rules are formulated based on the experience gained. To reduce the computational burden of defuzzification, the centroid method is selected to be the defuzzifier. After gaining a single output value, we round it to one of the five driving load levels.

For example, if the membership of the ith speed is μ_{S_i}, and that of acceleration is μ_{A_i}, then the firing strength μ_i of the premise is calculated based on the min operator. The firing strength of each rule is calculated as follows:

$$\mu_i = \min(\mu_{S_i}, \mu_{A_i}) \tag{6.24}$$

After the firing strength μ_i is calculated for all 25 rules, the corresponding driving load level is obtained by a weighted

Speed	Acceleration				
	NB	NS	Z	PS	PB
VL	VL	VL	VL	L	L
L	VL	L	L	L	M
M	L	M	M	M	H
H	M	M	H	H	VH
VH	H	H	VH	VH	VH

TABLE 6.11 Fuzzy Rules

average given by

$$L = \text{round}\left(\sum_{i=1}^{25} \mu_i \alpha_i \Big/ \sum_{i=1}^{25} \mu_i\right) \tag{6.25}$$

where α_i is a value of consequent parts on ith rule.

6.2.2 CNN-NDEKF-Based Forecasting

The goal is to decide the driving load level in a future time slot T_s based on the past driving sequence so that the value of the load level can be used in the HEV power control strategy to reduce the fuel consumption and emission. We address this problem as a pattern recognition problem, using CNN-NDEKF.

6.2.2.1 A Brief Introduction to CNN-NDEKF

In recent years, neural networks have shown great promise in identifying complex nonlinear mappings from observed data and have found many applications in nonlinear systems. Despite significant progress in the application of neural networks to many real-world problems, however, the vast majority of neural network research still relies on fixed-architecture networks trained through back propagation or some other slightly enhanced gradient descent algorithm. There are two main problems with this prevailing approach. First, the "appropriate" network architecture varies from application to application; yet, it is difficult to guess this architecture, the number of hidden units, and the number of layers a priori for a specific application without some trial and error. Even within the same application, functional complexity requirements can vary widely, as is the case, e.g., in modeling human tracking strategies from different individuals.[67] Second, the back propagation and other gradient descent techniques tend to converge rather slowly, often exhibit oscillatory behavior, and frequently converge to poor local minima.

Therefore, Nechyba and Xu[68] developed a new neural network learning architecture to counter these problems mentioned earlier. This neural network is a well-known flexible cascade neural network with node-decoupled extended Kalman filtering. Below, we briefly summarize the CNN-NDEKF training algorithm and why we selected this learning algorithm to forecast the future driving load level.

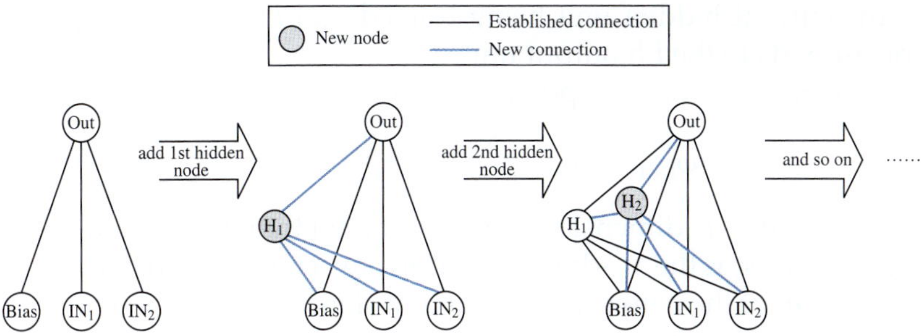

FIGURE 6.13 The cascade learning architecture.

First, no a priori model structure is assumed; the neural network automatically adds hidden units to an initially minimal network as the training requires. Figure 6.13 illustrates how a two-point, single-output network grows as two hidden units are added. Thus, a cascade network with inputs, hidden units, and outputs has connection where

$$\mathrm{n}_w = n_{\mathrm{in}} n_0 + n_h(n_{\mathrm{in}} + n_0) + (n_h - 1)\frac{n_h}{2} \tag{6.26}$$

Second, the hidden unit activation functions are not constrained to be a particular type. Rather, for each new hidden unit, the incremental learning algorithm can select that functional form which maximally reduces the residual error over the training data. Typical alternatives to the standard sigmoidal function are the sine, cosine, and gaussian functions.

Finally, it has been shown that NDEKF, a quadratically convergent alternative to slower gradient descent training algorithms, such as back propagation or quick propagation, fits well within the cascade learning framework and converges to good local minima with less computation. NDEKF is a natural formulation for cascade learning as we only train the input-side weights of one hidden neuron and the output units at any one time. We can partition the m weights by unit into groups—one group for the current hidden unit, and groups for the output units. In fact, by iteratively training one hidden unit at a time and then freezing that unit's weights, we minimize the potentially detrimental effect of the node decoupling.

Denote ω_k^i as the input-side weight vector of lengths m_i at iteration k, for unit $i \in 0, 1, \ldots, n_0$, where $i = 0$ corresponds to the current hidden unit being trained and $i \in 0, 1, \ldots, n_0$ corresponds to the ith output unit.

The NDEKF weight-update recursion is given by

$$\omega_{k+1}^{i} = \omega_{k}^{i} + (\psi_{k}^{i})^{T}(A_{k}\xi_{k})\phi_{k}^{i} \tag{6.27}$$

where ξ_k is the n_0-dimensional error vector for the current training pattern, ψ_k^i is the n_0-dimensional error vector for the partial derivatives of the network's output unit signals with respect to the ith unit's net input, and

$$\phi_{k}^{i} = P_{k}^{i}\zeta_{k}^{i} \tag{6.28}$$

$$A_{k} = \left\{ I + \sum_{i=0}^{n_0} [(\psi_{k}^{i})^{T}\phi_{k}^{i}]\,[\phi_{k}^{i}(\phi_{k}^{i})^{T}] \right\}^{-1} \tag{6.29}$$

$$P_{k+1}^{i} = (P_{k}^{i} - \phi_{k}^{i})^{T}(A_{k}\phi_{k}^{i})\phi_{k}^{i}(\phi_{k}^{i})^{T} + \eta I \tag{6.30}$$

where $(\psi_k^i)^T$ is the m_i-dimensional input vector for the ith unit, P_{k+1}^i is the $m_i \times m_i$ approximate conditional error covariance matrix for the ith unit, and η is a small real number which alleviates the singularity problem for P_{k+1}^i.

The flexible functional form that cascade learning allows is ideal for forecasting the future driving load levels. By making as few prior assumptions as possible in driving load forecasting, we improve the likelihood that the learning algorithm will converge to a good predictive model of the driving data.

The goal that we seek to achieve here is to forecast the future driving load level. We consider the driving load sequence as a measurable stochastic process and the knowledge behind it as an underlying stochastic process. A CNN-NDEKF is employed to generate a classifier for future load level identification, and the model parameters are updated through a learning process that ensures the model best represents the training data. The procedures for CNN-NDEKF-based learning can be summarized as follows:

1. Initially, there are no hidden units in the network, only direct input-output connections. These weights are trained first, thereby capturing any linear relationship between the inputs and the outputs.
2. With no further significant decrease in the root mean square (rms) error between the network outputs and

the training data (erms), a first hidden unit is added to the network from a pool of candidate units. These candidate units are trained independently and in parallel with different random initial weights by using the quick-propagation algorithm.

3. The best candidate unit will be selected and installed in the network if no more appreciable error reduction occurs; therefore, the first hidden node is produced.
4. Once the hidden unit is installed, the hidden-unit input weights are frozen, while the weights to the output unit are going to train again. This allows for such faster convergence of the weights during training than a standard multilayer feed forward network.
5. This process (from steps 2 to 4) is repeated until either the erms reduce sufficiently for the training set or the number of hidden units reaches a predefined maximum number.

In modeling the future driving load level, as with other poorly understood phenomena, we must rely on modeling by observation, or learning, rather than theoretical or physical derivation. A future driving load level is characterized by unique, complex, and unknown properties. So we require a learning paradigm that can cope with many difficult challenges. First, little, if anything, is known a priori about the structure, order, granularity, or delay. Second, the future driving load level is dynamic, stochastic, and nonlinear. It is prone to gradual changes over time. To address these challenges, the CNN-NDEKF mentioned earlier can satisfy the requirement by learning the driving load data.

6.2.2.2 Training Samples Buildup

To develop a load forecasting classifier, a CNN-NDEKF is selected because of its effectiveness in the classification of complex and nonlinearly separable target classes. A CNN-NDEKF classifies its input vector into one of five target load levels through a two-stage process.

To train the CNN network for the load forecasting classification problem, the statistics of three typical drive cycles, i.e., US06 (average speed: 77.9 km/h), NEDC (average speed: 33.6 km/h), and Manhattan (average speed: 11.0 km/h), will be calculated to generate the training database. These cycles represent a wide variety of conditions because no two cycles seem to provide similar features; they are significantly different in acceleration rate, and average speed.

The corresponding flow diagram to prepare training data is shown in Fig. 6.14, and the procedure for one driving cycle is described below:

1. Initialize the segment lengths of T_p and T_f. Here T_p represents a past time slot which is used to forecast the future load level, while T_f represents an objective the future time slot.
2. Randomly choose a start point T_s, If $T_s + T_p + T_f$ is larger than the driving cycle length or T_s is smaller than T_p, repeat step 2.
3. Apply the fast Fourier transform (FFT) on the past driving cycle to select characteristic features and generate the CNN-NDEKF input vector P.
4. Calculate the average speed and acceleration of the future driving cycle. Apply the fuzzy logic approach introduced in Sec. 6.2.1 to calculate the load level as CNN-NDEKF output y. Generate a training sample $\{P, y\}$
5. Go to step 2 until there are 100 training samples.

Since 3 driving cycles are considered here and 100 training samples are generated from each driving cycle, we can obtain 3×100 training samples totally.

6.2.3 Simulation Study

In this simulation, we try to forecast the future load level by analyzing the past load sequence.

6.2.3.1 Segments Selection

As mentioned earlier, the selection of T_p and T_f will influence the performance of training significantly. Generally speaking, a larger T_p means more past information will be considered so that the corresponding result will be better while the computational cost will be much greater. Similarly, a smaller T_f implies

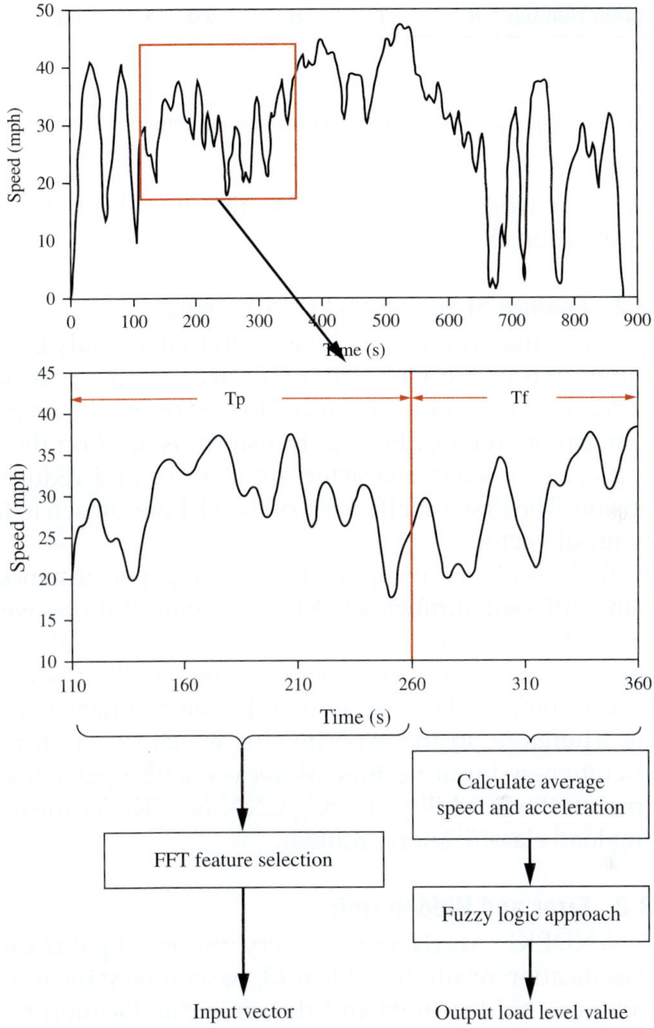

FIGURE 6.14 Flow of training data preparation.

that the classifier will work more frequently but cost more computation.

To balance the performance and cost, we let $T_p = 150$ and $T_f = 100$. The reason is that the typical cycle in urban traffic situations is approximately 180 s. So the value of T_p, i.e., 150 s, which is slightly less than 180 s, will be enough to reflect the past driving cycle. The value of T_f, i.e., 100 s, will make

Features Number	4	6	8	10	13	15	20
Error rate (%)	36.67	22.67	10.67	7.5	7.333	6.67	6

TABLE 6.12 Training Performance Comparison of Different Numbers of Coefficients

the classifier work in every cycle so that it will not miss any traffic situation changes.

6.2.3.2 Feature Selection and CNN Training

The past driving cycle cannot be used to train directly because the dimension of the past driving cycle, i.e., 150, is so large that it requires numerous instances to determine the result. In the application, the fast Fourier transform is used on the past driving cycle to select characteristic features and reduce the dimension. The first n coefficients of the FFT are chosen to form a new input vector.

In Table 6.12, we compare the training performance by choosing different numbers of FFT coefficients. It shows we can achieve better accuracy by choosing more coefficients, while the time consumption increases. It can be seen that when the number of coefficients is bigger than 10, the accuracy increases slowly. Therefore, in the experiments, we choose 10 features to generate new input vectors. Moreover, a 92.5 percent accuracy proves the feasibility of using CNN-NDEKF for the future driving load classification problem.

6.2.3.3 Error and Hidden Unit

In CNN-NDEKF, *MaxHidden* is a very important parameter to the classification result. In Table 6.13, we compare the number of maximum hidden units and the error classification rate of the CNN-NDEKF classifier with different *MaxHidden* values.

As shown in Table 6.13, when *MaxHidden* is 5, we get a minimum classification error rate of 7.5 percent. This result shows that the approach can get a high accuracy in the future driving

MaxHidden	2	4	5	6	8	10	15	20
Error rate (%)	26.67	13.33	7.5	17.33	22.67	23.33	23.33	24

TABLE 6.13 Number of Maximum Hidden Units versus Error Rate

Drive Cycle	HWFET	UDDS	WVUCITY5	WVUINTER	NYCC
Error rate (%)	11.33	8.67	11.33	16.67	17.33

TABLE 6.14 Training Performance Comparison of Different Numbers of Coefficients

load level classification. When *MaxHidden* increases from 2 to 5, the error rate reduces from 26.67 to 7.5 percent. Contrarily, it increases from 7.5 to 24 percent when *MaxHidden* increases from 5 to 20. A larger *MaxHidden* might correspond to higher testing accuracy, as well as more iterations; however, overfitting will occur. To avoid the overfitting, it cannot be too large. Here we choose the maximum hidden unit number to be 5.

6.2.3.4 Case Study

After obtaining the CNN-NDEKF classifier in the previous sections, we use five different standard driving cycles to evaluate it: highway fuel economy test (HWFET), urban dynamometer driving schedule (UDDS), city driving for a heavy vehicle (WVUCITY), interstate driving for a heavy vehicle (WVUINTER), and New York city cycle (NYCC). Obviously, these cycles represent a wide variety of conditions. It is effective to validate the classification performance by these driving cycles.

Simulation results of this case are shown in Table 6.14. It can be seen that an average error rate of 13.066 percent is gained, which proves the feasibility of the driving load forecasting approach.

6.3 Energy Management

As mentioned in Sec. 6.1, the main power control objective can be summarized as to satisfy the power requirement with the cooperation of the electric power source and fuel power source. Synchronously, the corresponding fuel consumption should be as low as possible.

For a hybrid electric vehicle, the fundamental problem is the optimal split of the powers of the combustion engine and the electric motor. It has to be solved "online" during the operation. Basically, the power flow of HEV can be represented by

$$P_d(t) = P_e(t) + P_b(t) \tag{6.31}$$

where the demand power $P_d(t)$ is a direct consequence of the vehicle based on the vehicle characteristics and speed, the driver demand, and the road situations. If we define the driving motor input direction as the positive axis, then it can be a positive, zero, or negative value because the motor can regenerate power when braking. To satisfy this demand power, the power control approach must determine the split of fuel source power P_e and battery power P_b and act to maximize the fuel efficiency and minimize the emissions. We define the generator output direction and battery output direction as the positive axes of P_e and P_b, respectively. Therefore, P_e will be nonnegative data, and P_b can be positive, zero, or negative.

The motivation of the the power control strategy is to determine the values of P_e and P_b so that Eq. (6.31) is satisfied based on the vehicle situation. Synchronously, the corresponding fuel consumption should be as low as possible.

In this section, first we treat this objective as an optimization problem and propose an approach based on model predictive control (MPC). Then we present a real-time MPC system, which combines the MPC controller and load forecasting system.

Figure 6.15 shows the overall architecture of the real-time MPC controller. Different from the simple MPC controller which will be discussed in Fig. 6.16 in Sec. 6.3.2.3, a "load

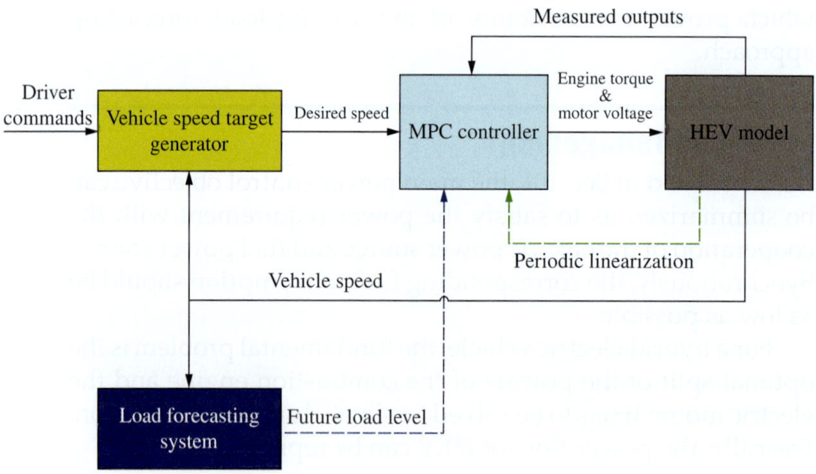

FIGURE 6.15 Real-time model predictive power control system architecture.

forecasting system" block is added to calculate the real-time future load level and sent it to the "MPC controller" block. Then, the MPC controller will modify weights based on future load level to obtain optimal performances. In this book, we use the vehicle speed target generator based on former research. See Schroedl and Zhang[97] for more information about it.

6.3.1 Performance Criteria Selection

The main reason for hybrid vehicle development is to make them better than conventional vehicles. In comparing different designs it is important to make it clear what performances should be compared and why these performances are chosen. Making an adequate performance index is complicated because it will almost certainly end in comparing quantities that are not comparable on an equal basis.

For a hybrid electric vehicle, several things have to be considered when selecting a performance index. All of them belong to one category such as operational costs or capital costs, while other things can belong to several categories. It is very hard to compare some factors, such as to compare or put a price on different kinds of noise, or if the vehicle behaves as a vehicle should behave (ease of handling). Consider these:

1. *Vehicle performance.* The first criterion we need to consider is the vehicle performance such as maximum speed (km/h), acceleration time (s), and gradeability (%). It is not acceptable to save energy by reducing the vehicle performance.
2. *Batteries lifetimes.* Batteries (the energy storage) are a vital part of a hybrid vehicle. They are initially a capital cost but will probably need replacement during the lifetime of the vehicle. The lifetime of the battery is heavily dependent on how it is charged and discharged which in turn depends on the size and type of battery, size and treatment of the ICE, and the power control strategy for the vehicle. The lifetime of batteries expressed in number of charge-discharge cycles can vary from several thousand cycles to less than 100 cycles. Generally speaking, keeping batteries in a proper level range can distinctly improve the batteries' life.

3. *Fuel consumption*. Fuel is naturally a running cost. Reducing fuel consumption is the main purpose of HEVs.

6.3.2 Model Predictive Control Approach

6.3.2.1 Problem Definition

As shown in Fig. 6.10, the vehicle system is a two-input eight-output system. The control objective is to determine, at each sampling time, the optimal pair of input variables (T_e and V_m) according to the minimization of the total fuel consumption over the speed cycle.

$$\min_{T_e(k),V_m(k)} J(k) = \sum_{i=0}^{N} \{w_1(V_x(k+i) - V_{xd}(k+i)]^2 + w_2\left[F_{\text{rate}}(k+i) - F_{\text{rate,max}}\right]^2\} \tag{6.32}$$

under the constraints

$$\begin{aligned} &0 \leq T_e(k) \leq T_{e,\max} \\ &V_{m,\min} \leq V_m(k) \leq V_{m,\max} \\ &0 \leq V_x(k) \leq V_{x,\max} \\ &0 \leq F_{\text{rate}}(k) \leq F_{\text{rate,max}} \\ &w_1 \geq 0 \\ &w_2 \geq 0 \end{aligned} \tag{6.33}$$

where $T_e(k)$ is the engine output torque; $T_{e,\max}$ is the engine maximum torque at time k; $V_m(k)$ is the driving motor torque; $V_{m,\min}$ is the minimum workable torque of the driving motor; $V_{m,\max}$ is the maximum torque the driving motor supply; $V_x(k)$ is the vehicle speed; $V_{x,\max}$ is the maximum speed the vehicle can achieve; $V_{xd}(k)$ is the desired vehicle speed; $F_{\text{rate}}(k)$ is the fuel rate (efficiency) of engine; $F_{\text{rate,max}}$ is the maximum fuel rate (efficiency) of engine; and w_1 and w_2 are weighting factors for vehicle speed and fuel rate, respectively, which should be nonnegative.

Equations (6.32) and (6.33) can be explained as follows: At time k, we want to obtain a limited pair of engine output torque and motor torque supplied on the driving motor. By applying them, we can make the vehicle follow a desired speed and can maximize the engine efficiency synchronously. Different choices of weighting factors will achieve different tradeoffs.

Obviously, this optimization problem has a solution corresponding to $T_e(k) = 0$, which means the vehicle operates purely electrically. This mode will use only electric power and, of course, lead to the battery discharge. To avoid this situation, a constraint on the battery state of charge should be introduced into the problem. For the sake of convenience, the following criterion is considered:

$$[\mathrm{SOC}(k) - \mathrm{SOC}_d(k)]^2 = 0 \tag{6.34}$$

where $\mathrm{SOC}(k)$ and $\mathrm{SOC}_d(k)$ are the battery SOC and the desired SOC at time k. A special case is that when the desired SOC, $\mathrm{SOC}_d(k)$, equals the initial SOC, $\mathrm{SOC}_{\mathrm{init}}$, $\delta\mathrm{SOC} = \mathrm{SOC}(k) - \mathrm{SOC}_d(k)$ presents the electric energy consumption over the speed cycle. If $\delta\,\mathrm{SOC} = 0$, we can consider that, globally speaking, the total amount of energy for propelling the vehicle has been taken from the irreversible source of fuel. In this case, the fuel consumption results can be compared with those of conventional vehicles.

During the operation, we need the vehicle to run purely electrically. To achieve this situation, we also consider the engine speed as one of the criteria. By controlling the engine speed to zero, the MPC controller can make the vehicle run purely electrically.

$$[E_s(k) - E_{\mathrm{SD}}(k)]^2 = 0 \tag{6.35}$$

where, $E_s(k)$ and $E_{\mathrm{SD}}(k)$ are engine speed and the desired engine speed at time k.

Another criterion is the control variable's rate. A sudden and large change of the engine torque or driving motor torque may cause the vehicle to be unstable. Therefore, the changing rates of the input variables also need to be considered.

Then, the global optimization problem is transformed to the following equation, under the constraints in Eqs. (6.37).

$$\min_{T_e(k),V_m(k)} J(k) = \sum_{i=0}^{N} \left\{ \begin{array}{l} w_1[E_s(k+i) - E_{SD}(k+i)]^2 \\ +w_2[V_x(k+i) - V_{xd}(k+i)]^2 \\ +w_3[\mathrm{SOC}(k+i) - \mathrm{SOC}_d(k+i)]^2 \\ +w_4[F_{rate}(k+i) - F_{rate,max}]^2 \end{array} \right\}$$

$$+\sum_{i=0}^{N_u} \left[w_5 \Delta T_e(k+i)^2 + w_6 \Delta V_m(k+i)^2 \right] \tag{6.36}$$

Constraints:

$$\begin{aligned} 0 \le E_s(k) &\le E_{s,max} \\ 0 \le T_e(k) &\le T_{e,max} \\ V_{m,min} \le V_m(k) &\le V_{m,max} \\ 0 \le V_x(k) &\le V_{x,max} \\ 0 \le F_{rate}(k) &\le F_{rate,max} \\ \mathrm{SOC}_{min} \le \mathrm{SOC}(k) &\le \mathrm{SOC}_{max} \\ \Delta T_{e,min} \le \Delta T_e(k) &\le \Delta T_{e,max} \\ \Delta V_{m,max} \le \Delta V_m(k) &\le \Delta V_{m,max} \\ w_1 &\ge 0 \\ w_2 &\ge 0 \\ w_3 &\ge 0 \\ w_4 &\ge 0 \\ w_5 &\ge 0 \\ w_6 &\ge 0 \end{aligned} \tag{6.37}$$

where SOC_{min} and SOC_{max}, respectively, are the minimum and maximum operational battery SOC level. In Sec. 6.1.1.3, we have explained why the battery operational power level should be predefined. Both $\Delta T_{e,min}$ and $\Delta T_{e,max}$ define the range of engine torque variation every time; $\Delta V_{m,min}$ and $\Delta V_{m,max}$ define that of the driving motor.

6.3.2.2 A Brief Introduction of Model Predictive Control

An attractive solution for the earlier optimization problem is model predictive control, which can solve the problem with constraints of the inputs, outputs, and states as well. MPC had been widely used during the past years in industry.[10, 58, 93]

Before applying the MPC on HEV power control problem, we present a brief introduction to MPC.

Consider a constrained linear discrete-time system

$$\begin{aligned} x(k+1) &= Ax(k) + Bu(k) \\ y(k) &= Cx(k) + Du(k) \\ x(0) &= x_0 \end{aligned} \tag{6.38}$$

with the control constraints

$$\begin{aligned} &y(i)_{\min} \le y(i) \le y(i)_{\max} \qquad i = 1, 2, \ldots, m \\ &u(j)_{\min} \le u(j) \le u(j)_{\max} \\ &\Delta u(j)_{\min} \le \Delta u(j) \le \Delta u(j)_{\max} \qquad j = 1, 2, \ldots, n \end{aligned} \tag{6.39}$$

where k is the discrete time index; $x \in R^p$ represents the system state; $u \in R^n$ is the input vector; $y \in R^m$ is the output vector, $y(i)_{\min}$ and $y(i)_{\max}$, $i = 1, 2, \ldots, m$, are the output constraints; $u(j)_{\min}$ and $u(j)_{\max}$, $j = 1, 2, \ldots, n$, are the input constraints; and $\delta u(j)_{\min}$ and $\delta u(j)_{\max}$, $j = 1, 2, \ldots, m$, are the output rate constraints.

Model predictive control is, at time instant k, to find a control sequence $u(k|k), \ldots u(k+N-1|k)$ so that a specified performance index $J(k)$, typically defined Eq. (6.40), in the moving horizon time frame is minimized and the constraints Eq. (6.39) are satisfied.

$$\begin{aligned} J(k) = &\sum_{i=0}^{N} [\hat{y}(k+i|k) - r(k+i|k)]^T Q [\hat{y}(k+i|k) - r(k+i|k)] \\ &+ \sum_{j=0}^{N_u} \Delta\hat{u}(k+j|k)^T R \Delta\hat{u}(k+j|k) \end{aligned} \tag{6.40}$$

where $\hat{y}(k+i|k)$ is the predictive output vector at time k; $r(k+i|k)$ is the reference/desired trajectory vector; $\Delta\hat{u}(k+j|k)$ are the changes of the future input vector; $Q \ge 0$ and $R \ge 0$ are output and input rate weighting matrices, respectively; N

is the length of the predictive horizon; and N_u is the length of the predictive control horizon.

The MPC problem to be discussed is formulated mathematically as follows subject to Eqs. (6.38) and (6.39).

$$\min_{u(k|k),\ldots,u(k+N-1|k)} J(k) \tag{6.41}$$

The problem in Eq. (6.41) is solved online using an optimization solver, e.g., quadratic programming (QP), which yields a control sequence $u(k|k), \ldots u(k+N-1|k)$. The MPC law is then determined by

$$u(k) = u(k|k) \tag{6.42}$$

If the control effort in each time instant is given by

$$u(k+j|k) = K(k+j)x(k), \qquad j = 0, 1, \ldots, N_u - 1 \tag{6.43}$$

where $K(k+i)$ is the control gain at the time instant $k+i$, then the predicted state $x(k+N|k)$ at time instant $k+N$ driven by the control sequence $u(k|k), \ldots, u(k+N-1|k)$ from the state $x(k)$ can be expressed by

$$x(k+N|k) = \left[A^N + \Gamma K_N(k)\right] \cdot x(k) \tag{6.44}$$

where

$$\begin{aligned} \Gamma &= \left[A^{N-1}B \quad \cdots \quad A^0 B\right] \\ K_N(k) &= \begin{bmatrix} K(k) \\ \vdots \\ K(k+N-1) \end{bmatrix} \end{aligned} \tag{6.45}$$

6.3.2.3 Model Predictive Power Control Method for HEV

After a summary of the hybrid electric power control problem and a brief introduction of MPC, we introduce an HEV power control method based on MPC.

The overall control scheme is shown in Fig. 6.16. It is composed of two parts: a module which collects driver's commands

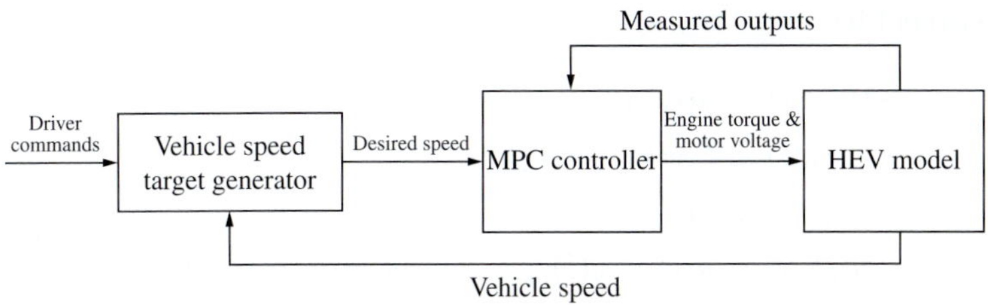

FIGURE 6.16 Overall HEV power control scheme. Here we focus on the design of the MPC controller.

(acceleration pedal and brake pedal commands), and consequently generates a desired vehicle speed r_s, and a MPC controller that regulates the torque from the engine and motor at the desired vehicle speed r_s. Here we focus on the second part; thus, we assume the desired vehicle speed is known and validate the MPC power controller by standard test driving cycles and simulation results.

The MPC controller receives the desired vehicle speed r_s and the measured *Engine Speed*, *EG-Current*, V_x, *Kappa*, *Motor Speed*, *Motor Current*, SOC, *Fuel eff* as model outputs and then calculates the desired engine torque T_e and motor torque V_m. Then the engine controller and motor controller receive them as set points and consequently control the engine and motor to the desired point.

In the following, we describe how a MPC controller can be designed for the hybrid electric vehicle power problem. The main idea of MPC is to use the model of the vehicle to predict the future evolution of the system. Based on the prediction, at each time step k, a certain performance index (cost function) is optimized under operating constraints with respect to a sequence of future input moves. The first of the optimal moves is the control action applied to the vehicle at time k. At time $k+1$, a new optimization is solved over the shifted horizon. For the HEV power control problem, the following optimal problem is solved:

$$J(k) = \sum_{i=0}^{N} [\hat{y}(k+i|k) - r(k+i|k)]^T Q [\hat{y}(k+i|k) - r(k+i|k)] + \sum_{j=0}^{N_u} \Delta\hat{u}(k+j|k)^T R \Delta\hat{u}(k+j|k) \tag{6.46}$$

subject to

$$
\begin{aligned}
x(k+1) &= Ax(k) + Bu(k) \\
y(k) &= Cx(k) + Du(k) \\
x(0) &= x(k) \\
y(i)_{\min} &\le y(i) \le y(i)_{\max} \qquad i = 1, 2, \ldots, m \\
u(j)_{\min} &\le u(j) \le u(j)_{\max} \\
\Delta u(j)_{\min} &\le \Delta u(j) \le \Delta u(j)_{\max} \quad j = 1, 2, \ldots, n
\end{aligned}
\tag{6.47}
$$

where Q and R are the output and input rate weighting matrices, respectively. We have explained why we need to measure outputs more than the performance criterion. Thus, $Q = \mathrm{diag}(w_1, 0, w_2, 0, 0, 0, w_3, w_4)$ and $R = \mathrm{diag}(w_5, w_6)$, and we just consider the elements that will affect performance and let other elements be zero. Because the ranges of outputs are remarkably different (0 to 6000 for engine speed, 0 to 150 for vehicle speed, 0.4 to 0.9 for SOC, and 0 to 0.34 for fuel efficiency), we need to consider the normalization for outputs to balance their contributions for the cost function. In practice, the weights are bounded as $0 \le w_1 \le 1/6000$, $0 \le w_2 \le 1/150$, $0 \le w_3 \le 1/0.5$, and $0 \le w_4 \le 1/0.34$. Also N is the length of the predictive horizon, and N_u is the length of the predictive control horizon.

The parameters in the MPC controller Eqs. (6.46) and (6.47) to be modified are the prediction horizon length N, the predictive control horizon N_u, and weight matrices Q and R.

By increasing the prediction and control horizon N and N_u, the controller performance improves, but the computational cost increases at the same time. So, tuning N and N_u to find smaller values that lead to a satisfactory performance is important. As mentioned in Sec. 6.1.3.3, we cannot control all performance criteria to follow the reference trajectory exactly. Thus, the choice of weighting factors is crucial. They will affect the balance between good tracking and acceptable performance.

6.3.3 Particle Swarm Optimization

The complex interaction between the various MPC weights makes it difficult to determine specific weights manually or analytically. So simulation-based optimization is a possible solution for these kinds of system level optimization problems.

Optimization tries to minimize fuel consumption by searching the multidimensional weight space for the various combinations of operation weights (optimization variables) and selecting the best combination at each iteration. In other words, it eliminates the bad operation weights while keeping the good ones. The goal is to select appropriate parameters (operational weights) so that the fuel consumption is reduced after a certain load level drive cycle. We formulate the goal as an optimization problem.

Particle swarm optimization (PSO), discovered through simulation of the social behavior of a bird block, was first proposed by Eberhart and Kennedy[20, 44] and used for optimization of continuous nonlinear functions. The swarm consists of some volumeless particles with velocities, each of which represents a feasible solution in the solution space. The PSO algorithm finds the optimal solution by moving the particles in the solution space. Besides the PSO for continuous nonlinear optimization problems, two novel PSO algorithms were developed to solve various problems. Kennedy and Eberhart[45] developed a discrete binary version of the particle swarm algorithm to solve combinatorial problems, and Yoshida et al.[119] proposed a new PSO to solve a mixed-integer nonlinear problem (MINLP).

Because of the advantages of realization and the promising optimization ability in various problems, PSO has been paid greater attention by researchers. Most of the works focus on two aspects. The first one is to improve the performance of PSO by modifying parameters, increasing the population diversity and hybrid with other optimizing approaches.[98],[15]. The second one is to apply PSO in different areas, such as multiobjective optimization,[16] electronics,[30, 119] training neural network,[111] medicine and emergent system identification,[112] etc.

After studying the optimization methods for HEV power control strategy, with the progress of PSO in real-code optimization application, PSO is particularly suitable for the optimization of MPC operational weights.

6.3.3.1 Fitness Function and Optimization Variables

Optimal MPC weights can be defined as the solution for a constrained nonlinear programming problem as described by

$$\begin{aligned} &\min f(w_i) \\ &\text{such that } w_{\max} \leq w_i \leq w_{\min} \qquad i = 1, 2, 3, 4 \end{aligned} \tag{6.48}$$

where w_i, a suite of weights for power control strategy, is one solution for the nonlinear programming problem; $w_{\max} \leq w_i \leq w_{\min}$ is a group of inequality constraints; and $f(x_i)$ is the fitness function for power control strategy optimization. It presents the fuel consumption after a drive cycle.

For variables to optimize, w_1, w_2, w_3, and w_4, are the weights for EngineSpeed, VehicleSpeed, SOC, and Fuel effs, respectively. However, because we always have to follow the desired speed trajectory, w_2 is fixed to 1/150. The reason why we choose 1/150 as w_2 is that the vehicle maximum speed is 150 km/h. Thus 1/150 can normalize the contribution of vehicle speed to cost function. Therefore, only three optimization variables should be optimized, and similarly their bounds are $0 \leq w_1 \leq 1/6000$, $0 \leq w_3 \leq 1/0.5$, and $0 \leq w_4 \leq 1/0.34$.

6.3.3.2 PSO Algorithm Description

Particle swarm optimization is a population-based optimization procedure. It starts by initializing random solutions called *particles* in the multidimensional parameter space. In a PSO system, each particle flies in the multidimensional parameter space to look for the global minimum. Each particle in the PSO is defined by a point in the parameter space, called *position*, and its flight speed, called *velocity*. Also each particle is aware of its best position reached so far (*pbest*) and the best position of the group so far (*gbest*). During flight, each particle adjusts its position according to its own experience (*pbest* value) and according to the experience of its neighboring particles (*gbest* value). The position is modified using the concept of velocity. The velocity of each particle is updated as

$$v_i^{n+1} = kv_i^n + c_1 rand_1(pbest_i - p_i^n) + c_2 rand_2(gbest - p_i^n) \quad (6.49)$$

where v_i^{n+1} is the velocity of particle i at iteration $n+1$; k is the weighing function; c_1 and c_2 are the weighing factors; $rand_1$ and $rand_2$ are two random numbers between 0 and 1; p_i^n is the position of particle i at iteration n; $pbest_i$ is the best position particle i; and $gbest$ is the best position of the group (best of all $pbest$). Here $k = 0.7$ and $c_1 = c_2 = 2.1$ are chosen for better performance.[108] Similarly, the position is updated as

$$p_i^{n+1} = p_i^n + v_i^{n+1} \quad (6.50)$$

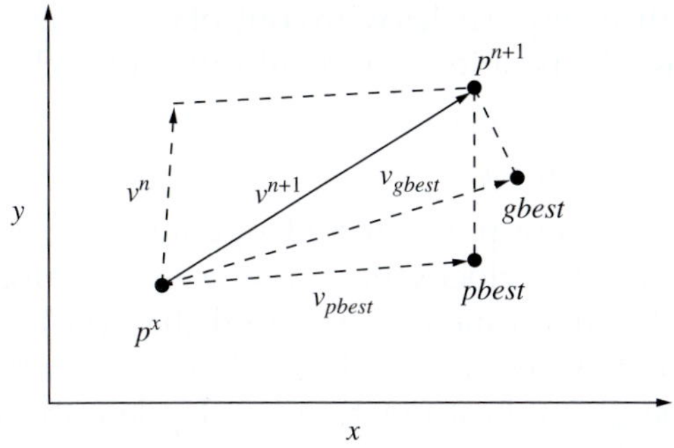

FIGURE 6.17 Updating velocity and position in PSO.

The velocity and position updating for particle i is illustrated in Fig. 6.17. Note that Fig. 6.17 represents a two-dimensional problem. It can be seen in Fig. 6.17 that the position of the particle is adjusting itself toward the *gbest* position. This is so because the velocity has changed its direction toward the *pbest* and *gbest* values.

6.3.3.3 Procedure Cycle

The first work to operate the PSO algorithm for the HEV parameter optimization problem is to choose the appropriate *particles*, which represents the potential solution to power control parameters. Based on the analysis of the optimization variables, we define the particles with the following coefficients:

$$(w_1, w_3, w_4)$$

6.3.3.4 Fitness Function

Once the particles have been defined, we need to define a fitness function to evaluate the status of each individual particle during the search. The fitness function of the corresponding particle is fuel consumption $f(x_i)$ after the drive cycle. For each individual particle, the fuel consumption is calculated, over a specific level drive cycle.

6.3.3.5 Computation Procedure

According to the PSO algorithm, the computation procedures to optimize weights on the specific cycles are as follows:

1. Initialize the particles with control vector size. Randomly choose the position p and velocity v of each particle.
2. Evaluate the particles by the objective function; find the best position x_{gb}.
3. Evaluate new position and update velocity. Move the particles by velocity v, $x_{\text{new}} = x + v$; evaluate x_{new} by the objective function; and find the best local position of the new set of particle x_{lb}. If $x_{gb} \geq x_{lb}$, then x_{lb} will be the global best position x_{gb}. Update the velocity by $v_{\text{new}} = kv + c_1 r_1 (x_{lb} - x) + c_2 r_2 (x_{gb} - x)$, where c_1, and c_2 are tuning parameters; k is inertial weight, and r_1, r_2 are random numbers from 0 to 1.
4. Repeat step 3 until the best position is sufficiently good or the maximum number of iterations is reached.

Here $k < 1$ is known as the *inertial weight* and it is chosen between 0 and 1 to determine to what extent the particle remains along its original course unaffected by the pull of the other two terms. These are balanced by the scaling factors c and r that are random, positive numbers with a uniform distribution and a value that goes from 0 to 1.

Therefore individual particles are attracted by both the position of their best past performance and the position of the global best performance of the whole swarm. Both deterministic and stochastic update rules are taken into consideration to update velocities; the algorithm calculates the next positions and moves all the agents following the new directions.

6.3.3.6 Results of Optimal Weights

The optimization process is executed for different load level drive cycles. As an example, we choose a $MediumLevel$ 100 s drive cycle. The drive cycle is a part of US06 drive cycle, which consists of a low-speed city drive cycle and high-speed highway drive cycle. The speed schedule of chosen $MediumLevel$ drive cycle is shown in Fig. 6.18.

The training process of fitness function (fuel consumption) based on particle swarm optimization for the above medium-level drive cycle is presented in Fig. 6.19. After 670 iterations, the fitness function evaluation converges to 29.9183.

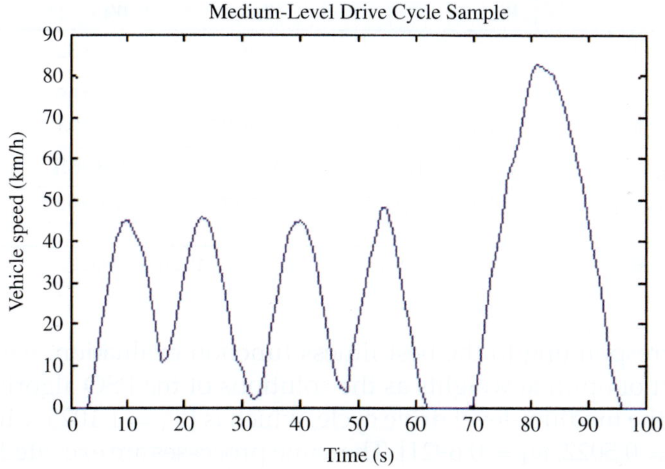

FIGURE 6.18 Medium-level drive cycle sample.

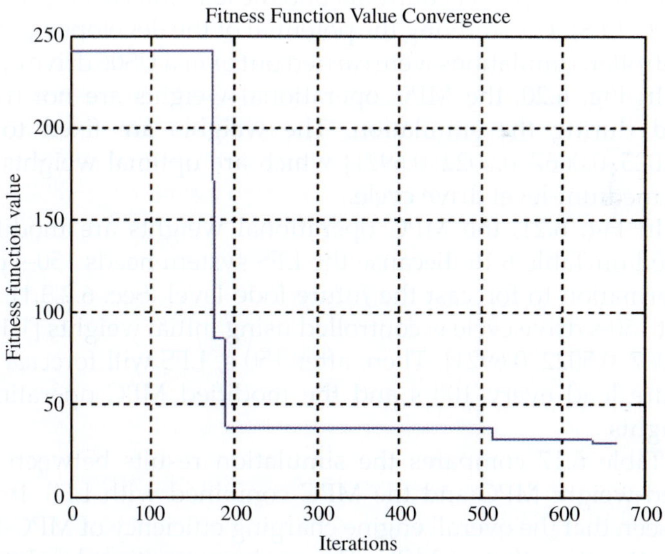

FIGURE 6.19 Particle swarm optimization results for medium-level drive cycle.

	w_1	w_3	w_4
Very low	1.3547×10^{-4}	1.065	1.4332
Low	3.2867×10^{-5}	1.246	2.8488
Medium	1.1025×10^{-4}	0.5022	0.6921
High	1.001×10^{-4}	1.1424	1.8065
Very high	1.9233×10^{-5}	0.5372	0.1294

TABLE 6.15 Optimal Weights for Different Load Level Drive Cycles

Corresponding to the best fitness function evaluation, we get a set of optimal weights as the solutions of the PSO algorithm for the medium-level drive cycle, which is [$w_1 = 1.1025 \times 10^{-4}$, $w_3 = 0.5022$, $w_4 = 0.6921$]. The same processes are executed for other load level drive cycles. The optimal weights are listed in Table 6.15.

6.3.3.7 Simulation Study

After obtaining the optimal weights for all load levels, we integrate the MPC power controller and the load forecasting system (MPC-LFS). For studying the potential of the developed power controller, simulations were carried out over a US06 drive cycle.

In Fig. 6.20, the MPC operational weights are not modified during the simulation. The weights are fixed to be [1.1025 0.0067 0.5022 0.6921] which are optimal weights for the medium-level drive cycle.

In Fig. 6.21, the MPC operational weights are modified based on Table 6.16. Because the LFS system needs 150-s past information to forecast the future lode level (Sec. 6.2.3.1), the first 150-s drive cycle is controlled using initial weights [1.1025 0.0067 0.5022 0.6921]. Then, after 150 s, LFS will forecast the future load every 100 s and the modified MPC operational weights.

Table 6.17 compares the simulation results between the fixed-weight MPC and the MPC combined with LFS. It can be seen that the overall engine charging efficiency of MPC-LFS is better than that of MPC. The engine operational points of MPC-LFS are more stable than those of MPC. Moreover, MPC-LFS can improve fuel consumption by 9.45 percent more than MPC.

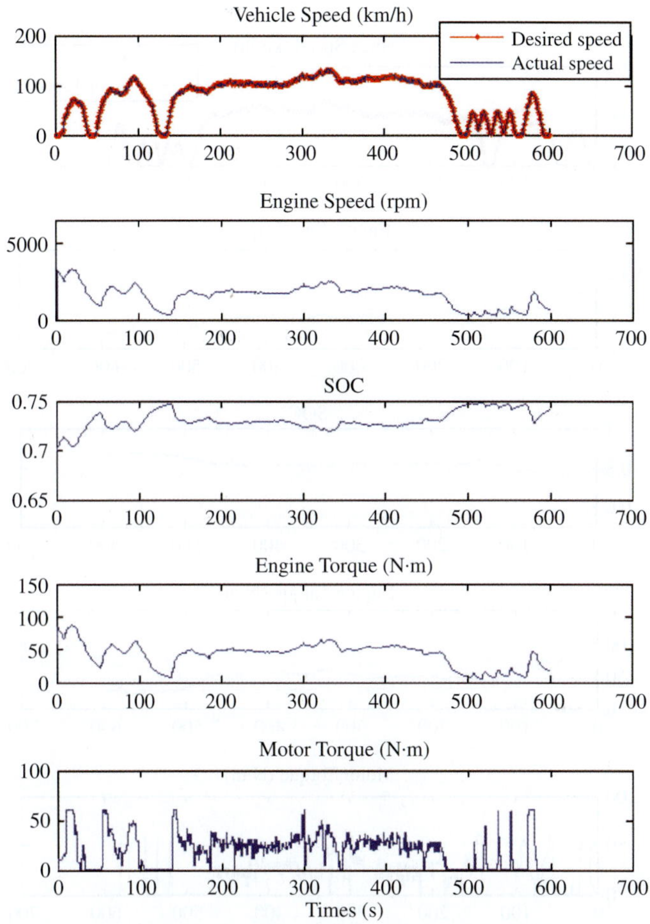

FIGURE 6.20 Simulation results over US06 drive cycle: fixed weights = [1.1025 0.0067 0.5022 0.6921].

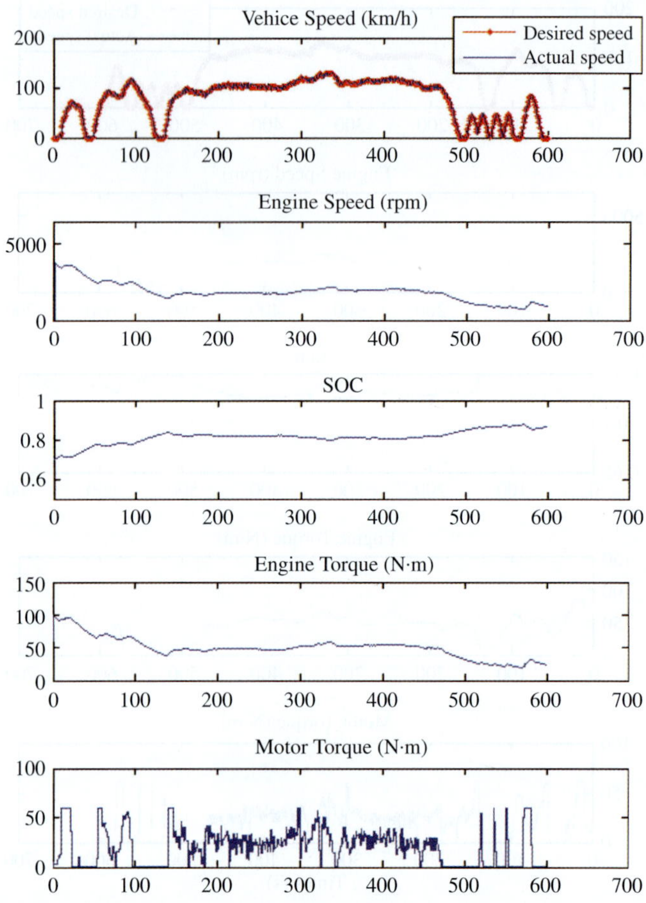

FIGURE 6.21 Simulation results over US06 drive cycle: MPC-LFS.

Time (s)	Load Level	Weights $w_1 \times 10^{-4}$	w_2	w_3	w_4
1 ∼ 150	M	1.1025	0.0067	0.5022	0.6921
151 ∼ 250	VH	1.9233	0.0067	0.5372	0.1294
251 ∼ 350	VH	1.9233	0.0067	0.5372	0.1294
351 ∼ 450	H	1.001	0.0067	1.1424	1.8065
451 ∼ 550	M	1.1025	0.0067	0.5022	0.6921
451 ∼ 601	M	1.1025	0.0067	0.5022	0.6921

TABLE 6.16 Weight Schedule for US06 Drive Cycle

	η_{eng}	ΔSOC	E_k	F_1	F_2	T_fuel
MPC	0.2462	0.0416	168.7145	375.3652	72.3838	302.9814
MPC+FLS	0.2629	0.1671	677.5559	546.6336	272.2940	274.3397
Fuel consumption improvement (%)						+9.45

TABLE 6.17 Simulation Results Comparison between MPC and MPC-LFS

6.4 Experiments and Analysis

This section describes the experiments that were carried out to verify the power control strategy. Because the vehicle has no license for public roads, we tested the vehicle on a private road about 400 m long. We tested the vehicle in three different modes: purely electrical, thermostat control, and MPC-LFS. The following sections provide experimental results and analysis.

6.4.1 Pure Electrical Experiment

The vehicle runs purely electrically in the first experiment. The experimental results are shown in Figs. 6.22 and 6.23.

Figure 6.22 shows the acceleration pedal position, vehicle speed, and battery temperature curve during the experiment. The acceleration pedal can output two different voltage ranges, 0.48 to 1.8 V and 0.96 to 3.6 V, for a total 30° pedal open angle. The road we test on is busy sometimes, so we choose the low-voltage range to test the vehicle for safety. In the figure, we normalize the 0.48- to 1.8-V voltage signal to 0 to 100 percent. The battery temperature is about 31°C. The BMS uses this temperature when calculating the SOC.

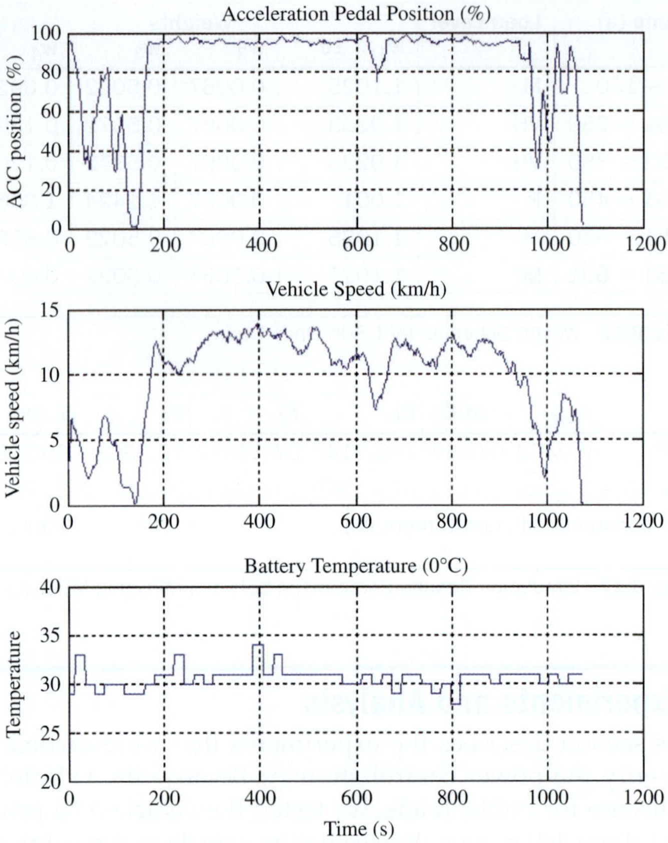

FIGURE 6.22 Pure electrical experiment results I.

Figure 6.23 shows the generator/battery current, bus voltage, and SOC of battery during the experiment. It can be seen in the figure that the battery output current is positive, which means the battery is discharging. The generator output current is zero because the engine is off. The bus voltage is between 48 and 50 V. Because the battery is discharging during the pure electrical mode, the SOC of the battery is decreasing. The final SOC is about 0.43.

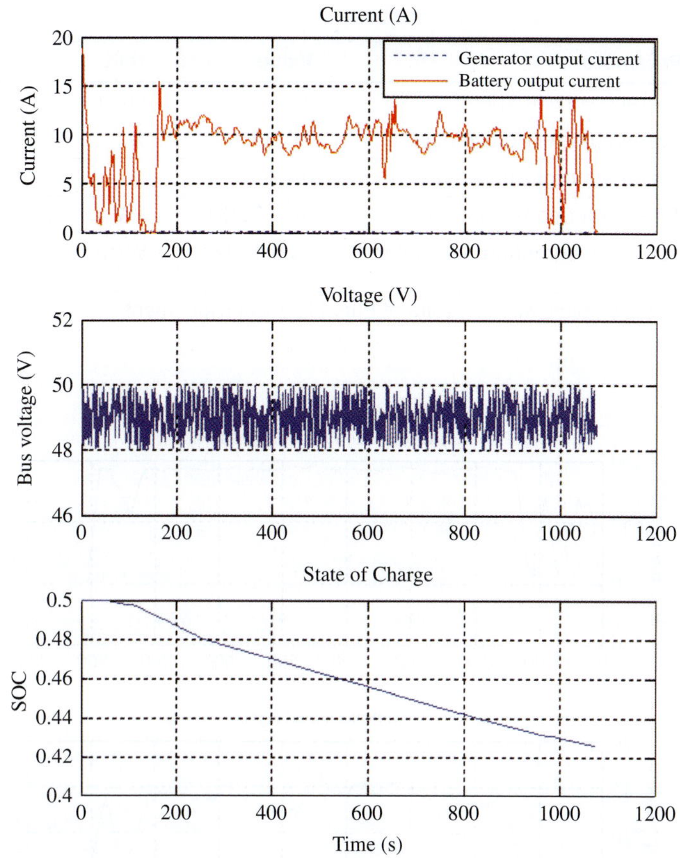

FIGURE 6.23 Pure electrical experiment results II.

Table 6.18 summarizes some postexperiment results. During the experiment, the vehicle's average speed is about 10.8298 km/h. The average efficiencies of the battery and motor are 80.38 and 84.14 percent, respectively. The vehicle consumes 462.0200 kW of electric power totally and 315.1018 W/m.

6.4.2 Thermostat Control Experiment

The second experiment is performed with engine power, and the thermostat control approach is used to control the vehicle. The experimental results are shown in Figs. 6.24 and 6.25.

Parameter	Value	Unit
Average vehicle speed	10.8298	km/h
Average battery efficiency	80.38	%
Average motor efficiency	84.14	%
Total power consumption	462.0200	kW
Power consumption per meter	315.1018	W/m

TABLE 6.18 Simulation Results of Pure Electrical Experiment

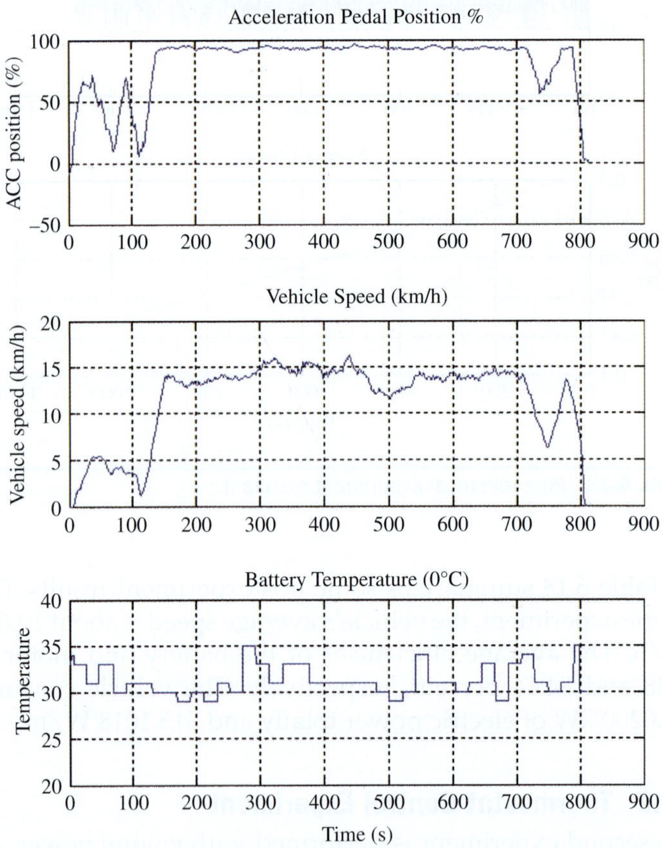

FIGURE 6.24 Thermostat control experiment results I.

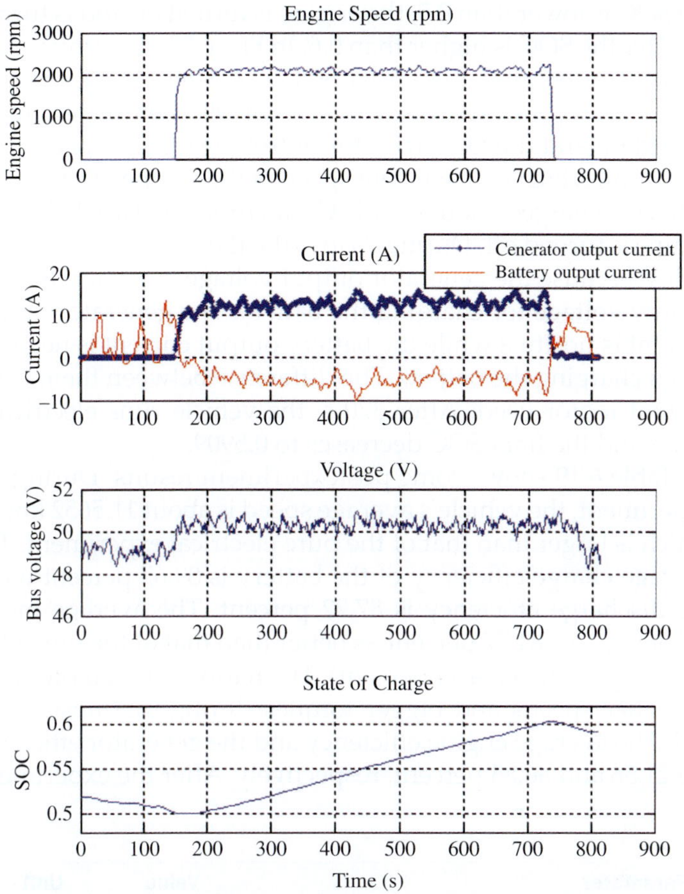

FIGURE 6.25 Thermostat control experiment results II.

Figure 6.24 shows the acceleration pedal position, vehicle speed, and battery temperature curve. Compared with the pure electrical experiment, during this experiment the average speed of the vehicle is higher. Since the driver is not a professional vehicle tester (although he intends to drive the same as he did in the pure electrical experiment), the drive cycle cannot be followed exactly.

Because the experimental road distance is limited, to reveal the basic philosophy of the thermostat control approach, we define a small operational SOC range as 0.5 to 0.6. Thus, when the SOC is lower than 0.5, the engine is turned on and is turned off after the SOC is higher than 0.6. In Fig. 6.25, the vehicle runs electrically because the initial SOC is 0.5185. At 75 s (x axis unit is 0.5 s), the SOC decreases to 0.4999; then the engine is turned on and the ECU tries to control the engine speed to be 2100 rpm, which is an efficient speed of the engine. At 370 s, the battery is charged enough and SOC increases to 0.6004; then the engine is turned off. During 75 to 370 s, the bus voltage is about 50.5 V because the generator output voltage is higher than the battery voltage for charging the battery. The generator output current is positive while the battery output current is negative when charging the battery. The difference between them is the current motor used. After 370 s, the vehicle runs electrically again and the final SOC decreases to 0.5909.

Table 6.19 shows some postexperiment results. During the experiment, the vehicle's average speed is about 11.5652 km/h, which is larger than that of the pure electrical experiment. The average charge efficiency of the battery is 89.67 percent while the discharge efficiency is 87.72 percent. The average motor efficiency, i.e., 84.77 percent, is better than that of the pure electrical experiment (84.14 percent). The reason is that motors run at higher speeds and bigger torques during thermostat control. The average engine efficiency and the generator efficiency are 21.56 and 86.85 percent, respectively. After the experiment,

Parameter	Value	Unit
Average vehicle speed	11.5652	km/h
Average battery charge efficiency	89.67	%
Average battery discharge efficiency	87.72	%
Average motor efficiency	84.77	%
Average engine efficiency	21.56	%
Average generator efficiency	86.85	%
SOC variations	0.0724	—
Fuel equivalent of electric energy variations	18.4224	g
Engine fuel consumption	87.5352	g
Total power consumption	69.1128	g
Power consumption per meter	0.0530	g/m

TABLE 6.19 Simulation Results of Thermostat Control Experiment

the end SOC is 0.0724 bigger than the initial SOC. The fuel equivalent of electric energy variations is about 18.4224 g under above-average engine and generator efficiencies. During the experiment, the engine consumes 87.5352 g of fuel. In the consumed fuel, 18.4224 g is stored in the battery. Then the total fuel consumption of the experiment is 69.1128 g, which means the vehicle uses 0.053 g of fuel per meter.

6.4.3 MPC-LFS Experiment

In the third experiment, the MPC-LFS controller is used. The proposed power control method is implemented with an IBM T60 laptop which communicates with ECU, BMS, and DSS by CAN bus.

Notes about the Experiment Here are some notes about the experiment:

- Because the vehicle is a series configuration, the motor expected torque is proportional to the acceleration pedal position.
- The road surface is assumed to be smooth, and there is no slide between the wheels and the road surface. Moreover, the pressure of the wheel is fully charged before the test. Under this condition, the kappa of the wheel is assumed to be a constant 0.014, which is a natural value for vehicles.
- The load forecasting algorithm operates in the experiment. However, for safety, the vehicle speed is limited to a low level and the future load levels are mostly the VL level.

Similarly, Fig. 6.26 shows the acceleration pedal position, vehicle speed, and battery temperature curve. Figure 6.27 shows the engine speed, current, bus voltage, and SOC information. At 20 s, the driver starts the vehicle. After about 9 s, the controller turns on the engine. At 450 s, the driver stops the vehicle. The engine is still on until we shut down the controller. During operation, the engine is controlled to a stable speed. The final SOC is 0.5957, which is 0.0557 bigger than the initial SOC (0.54).

Table 6.20 shows some postexperiment results. The vehicles's average speed is about 10.3188 km/h. The average charge

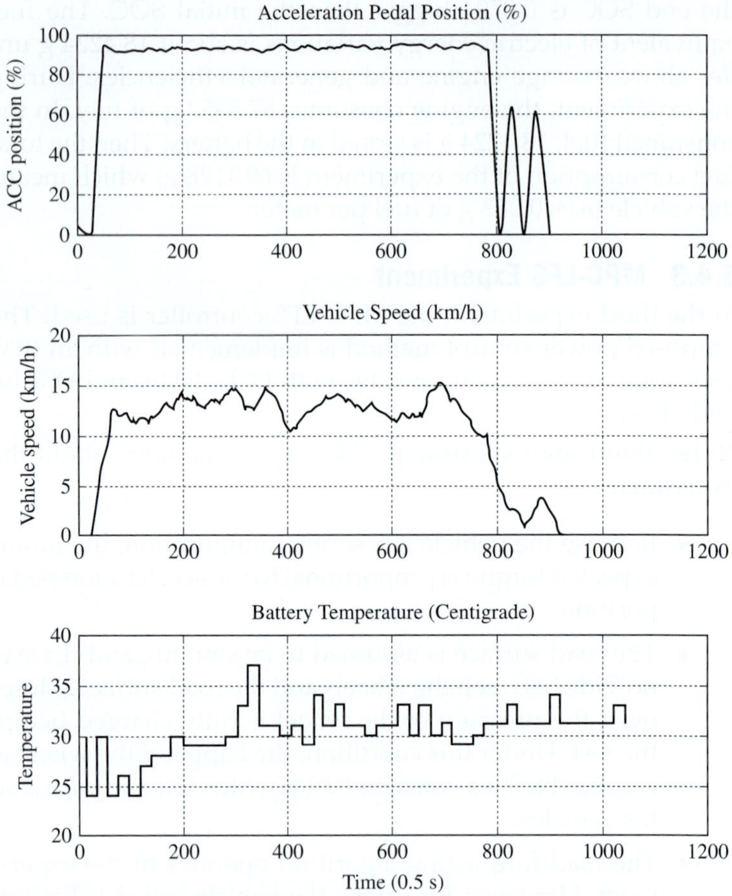

FIGURE 6.26 MPC-LFS experiment results I.

efficiency of the battery is 90.46 percent while the discharge efficiency is 89.57 percent. The average motor efficiency is 86.87 percent. The average engine efficiency and generator efficiency are 21.24 and 88.70 percent, respectively. The SOC variation is 0.0557. The fuel equivalent of electrical energy variation is about 13.1703 g. The engine consumes 74.2698 g of fuel. In the consumed fuel, 13.1703 g is stored in the battery for later use. In total, the fuel consumption of the experiment is 69.1128 g or 0.053 g of fuel per meter.

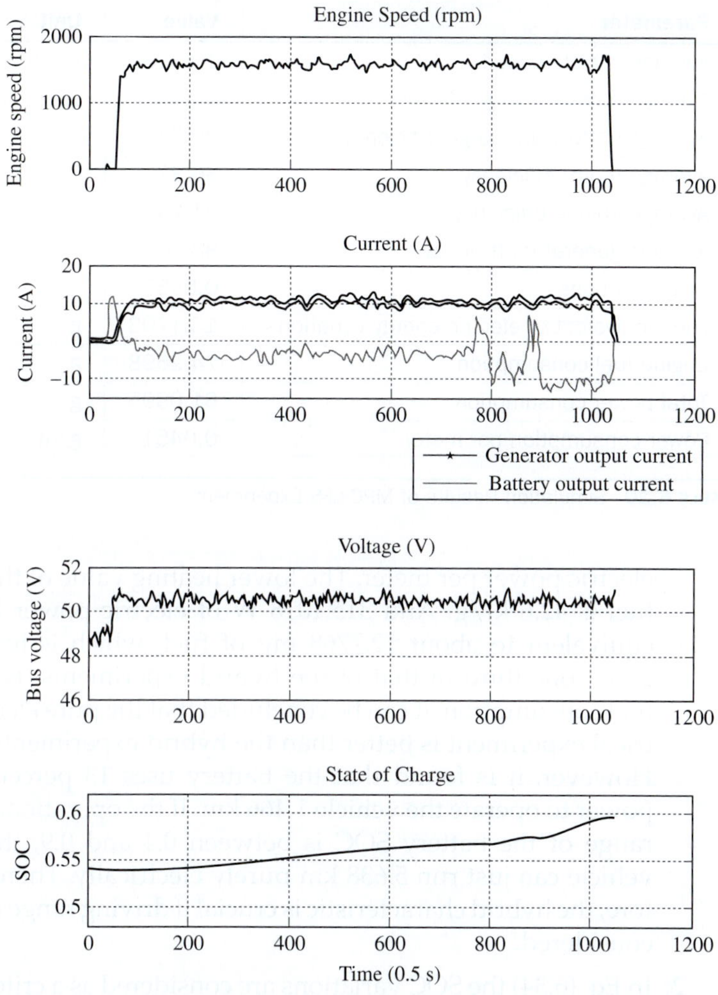

FIGURE 6.27 MPC-LFS experiment results II.

6.4.4 Comparisons

Through comparing the results of the three experiments, some phenomena are discovered:

1. Because the drive cycles in the three experiments are not exactly the same, we compare the power by fuel or power consumed every meter. It is found that in the pure electrical experiment, the vehicle consumes 103.79 W of

Parameter	Value	Unit
Average vehicle speed	10.3188	km/h
Average battery charge efficiency	90.46	%
Average battery discharge efficiency	89.57	%
Average motor efficiency	86.87	%
Average engine efficiency	21.24	%
Average generator efficiency	88.70	%
SOC variations	0.0557	—
Fuel equivalent of electric energy variations	13.1703	g
Engine fuel consumption	74.2698	g
Total power consumption	61.0995	g
Power consumption per meter	0.0451	g/m

TABLE 6.20 Simulation Results of MPC-LFS Experiment

electric power per meter. The lower heating value of the fuel is 42.6 kJ/g. And 315.1018 W of electric power is equivalent to about 12.7768 mg of fuel, which is just about one-third of that of the hybrid experiments. For fuel consumption, it can be concluded that the pure electrical experiment is better than the hybrid experiments. However, it is found that the battery uses 13 percent power to operate the vehicle 1.466 km. If the operational range of the battery SOC is between 0.4 and 0.9, the vehicle can just run 5.638 km purely electrically. Therefore, the hybrid characteristic is crucial if driving range is considered.

2. In Eq. (6.34) the SOC variations are considered as a criterion for optimization. Then the MPC-LFS will not allow the battery charges or discharges with large currents. It is well known that battery efficiency is low when charging or discharging large currents. The experiment results are consistent with this concept. The SOC variation in the MPC-LFS experiment is smaller than that in the thermostat control experiment, and both the battery charge and the discharge efficiency in the MPC-LFS experiment perform better.
3. Motor efficiencies in all experiments are almost the same because they are related to the drive's demand speeds.

4. Compared with the thermostat control experiment, the MPC-LFS experiment fuel consumption for 1 m is less. The reason is that the MPC-LFS considers the overall efficiency of the vehicle. It is found that although the engine efficiency in the MPC-LFS experiment is a little lower than that in the thermostat control experiment, the battery and generator efficiencies are improved. Generally speaking, the vehicle efficiency in the MPC-LFS experiment is better. The fuel consumption per meter improved by 14.9 percent from 0.0530 to 0.0451 g/m.

6.5 Summary

This chapter proposes a new power control approach for all kinds of HEVs. The models of key components of the series HEV are presented. The state's definition, linearization, and observability/controllability of the model are analyzed. In addition, the research of the load forecasting system and the simulation works are performed. Using fuzzy logic, the drive load is divided into five different levels. CNN-NDEKF is chosen as the classifier for its inherent advantages. Moreover, the method of MPC is presented and integrated with the load forecasting algorithm for the MPC-LFS for HEVs. Particle swarm optimization is used to find optimal operational weights for different drive levels. The MPC-LFS adjusts operational weights automatically to achieve the best performance. Finally, the implementation and experiments of the proposed MPC-LFS are completed. By analyzing the experimental results, the proposed approach is proved to be practical and efficient.

CHAPTER 7

Conclusions

This book covers a wide variety of research issues related to the latest technology of designing and controlling the omnidirectional HEV, from system design and steering control to intelligent battery and energy management. To conclude this book, the contributions to the field of omnidirectional HEV are outlined as follows.

7.1 Advancement in Vehicle Technology with State-of-the-Art Robotics Research

From the long years of study on wheeled mobile robots, the book proposes a novel design of an omnidirectional HEV by integrating the technology of omnidirectional motion of robots. The proposed HEV is capable of zero-radius turning which is impossible for traditional vehicles. While achieving the advantages of the conventional two-front-wheel steering vehicles, it can also realize additional functionalities such as ZRT and LP. In addition to proposing the modeling of the HEV, this book presents the implementation details and experimental results to illustrate the capability of the proposed HEV. Although some of the vehicle companies may have done research on omnidirectional vehicles, those technological documents are classified as confidential and are not released to the academic field. Hence, this book aims at providing a complete and self-contained documentation of the latest technological advancement on omnidirectional HEV for the researchers in academia.

7.2 Omnidirectional Vehicle Structure Supporting 4WIS and 4WID

In addition to the mechanical design of the HEV, the intelligent four-wheel independent steering and driving control mechanism is devised in this book. It involves two innovative elements.

First, an omnidirectional steer-by-wire system for the 4WIS vehicle, consisting of an extended steering interface and a behavior-based steering controller, is implemented. The proposed steering interface is an extension of the conventional steering interface which provides two more steering inputs so that the driver can steer the vehicle in either the conventional or omnidirectional way without any mode-switching operation. It is also able to provide feedback force so as to synchronize the extended steering interface and the orientations of the wheels. The behavior-based steering controller consists of the position controller and kinematic constraint controller. The kinematics constraint controller acts as a virtual linkage among the wheels to maintain kinematic constraint so as to minimize wheel slip. This approach is different from that in Refs. 106 and 35 which applied the trajectory tracking on each wheel. Since the driving path cannot be predefined by the control system for nonautonomous vehicles, the approach is beneficial as no trajectory planning is needed. The simulation results show that when compared with a position controller, the behavior-based controller improves both the driving and steering efficiency with a little path variation.

Second, the optimal traction force distribution and the deterministic yaw moment controller for the 4WIS-4WIS vehicle are proposed to enhance driving stability and controllability. The former approach does not require any feedback information, and no parameter tuning is needed. The latter stabilizes a vehicle by additionally tracking the heading angle of the vehicle instead of tracking a desired yaw rate as in the conventional method. The heading angle is obtained by a novel method based on centripetal force detection. It eliminates the prerequisite knowledge of the characteristics between wheels and road surface which are time varying and difficult to measure in real time. Moreover, it requires only low-cost sensing equipment such as the wheel speed sensor and accelerometer that makes the system practical to utilize. In general, the proposed method can be applied to any type of four-wheel independent drive vehicles.

7.3 Intelligent Battery Management System Specialized for HEV

Existing battery techniques cannot produce batteries that perfectly satisfy the requirements for application in HEV, especially the consistency of cells in the battery package and their safety in conditions of high-current or high-power charging and discharging. The proposed battery management scheme in this book is an effective improvement over the existing one in the following three aspects.

First, the existing typical SOC estimators such as coulomb counting and the extended Kalman filter cannot perform their theoretically optimal efficacy in practical applications due to various interference in the vehicle driving environment and model uncertainties. The proposed novel SOC estimator consists of (1) an adaptive-κ nonlinear diffusion filter to reduce the noise in current measurement, (2) a self-learning strategy to estimate and remove the zero drift, and (3) a coulomb counting algorithm to realize open-loop SOC estimators. The availability and efficacy of each component have been demonstrated based on comparative studies in simulation with the conventional approaches, under the testing conditions of noises with various SNR values, varying zero drifts, and different model errors. The overall framework has also been verified to rationally and efficiently combine these components and achieve robust estimation results in the presence of various kinds of noises and uncertainties.

Second, through the accurate estimation of SOC, the MPC-based fast-charging mechanism is proposed. It aims at simultaneously reducing the charge duration, which represents the out-of-service time of vehicles, and the increase in temperature, which represents safety and energy efficiency during the charge process. A single-mode lumped-parameter thermal model and a neural network trained by real experimental data are applied to predict the future temperature in simulations and experiments, respectively. A genetic algorithm is then applied to find the best charge sequence under a specified fitness function, which consists of two objectives: minimizing the charging duration and minimizing the increase in temperature. Both simulations and experiments demonstrate that the Pareto front of the proposed method dominates that of the most popular CCCV charge method.

Third, a fuzzy logic-based cell equalization controller is proposed and implemented, due to the lack electrochemical theories and models to explain the cell equalization process. The selection of inputs consists of (1) the difference in SOC, representing the degree of imbalance; (2) the average of SOC, representing the average discharge ability; and (3) total internal resistance, partially representing the equalizing efficiency. Simulation and experimental results both show that the proposed controller has the ability to incorporate the experiences and knowledge of experts, and intelligently tune the charging current along with the instantaneous battery states to balance the multiple objectives represented by the three inputs. Especially for $LiFePO_4$ battery, a large SOC imbalance may cause only a very small difference in voltage, leading to the failure of traditional voltage-based methods. Therefore, SOC-based equalization is more suitable for equalizing these types of batteries.

7.4 Intelligent Energy Management System Specialized for HEV

To meet the driver's power demand while optimizing fuel consumption and to manage the battery at a satisfactory power level to enable effective delivery of energy to the vehicle over a wide range of driving situations, the MPC-LFS-based power controller is able to solve some limitations of existing approaches, such as limited applicability, heavy computational cost, and difficult-to-achieve real-time performance.

There are two main contributions to this field in the book. First, a load level forecasting algorithm based on neural networks is developed. The key concept adopted in the development of the power control system is based on the premise that the future driving load would affect fuel consumption as well as the operating modes of the vehicles and driver behavior.

The proposed drive load forecasting algorithm is in charge of assessing the driving environment, the driving style of the driver, and the trend of the vehicles using long- and short-term statistical features of the past drive cycle.

Second, the MPC controller and the drive load forecasting algorithm are integrated, and a real-time control system is proposed. The system can be used for all kinds of HEVs based on the engine and the electric motor. A number of different performance indices have been applied to the control system. By changing the operational weights, the power control system can achieve different goals. Future load information is subsequently used by the power control system to change the operational weights of the performance, such as engine efficiency, engine speed, SOC, etc., which improves the vehicle's overall performance. In addition, particle swarm optimization is applied to find optimal operational weights for all drive levels. The improvement between the optimal and original controller is shown by experiments.

Abbreviations

2WS Two-wheel steering

4WD Four-wheel drive

4WID Four-wheel independent drive

4WIS Four-wheel independent steering

4WS Four-wheel steering

BLDC Brushless dc motor

BMS Battery management system

BP Back propagation

CCCV Constant-current, constant-voltage

CNN Cascade neural network

CNN-NDEKF Cascade neural network with node-decoupled extended Kalman filtering

DYC Direct yaw (moment) control

ECU Engine control unit

EM Electric motor

ESC Enhanced self-correcting

ESD Electrostatic discharge

EV Electric vehicle

FWS Front-wheel steering

GA Genetic algorithm

HEV Hybrid electric vehicle

ICE Internal combustion engine

ICR Instantaneous center of rotation

LP Lateral parking

MCU Microprogrammed control unit

MPC Model predictive control

NDEKF Node-decoupled extended Kalman filtering

NDF Nonlinear diffusion filter

OCV Open-circuit voltage

PWM Pulse-width modulation

RC Resistive-capacitative

SNR Signal-to-noise ratio

SOC State of charge

SPI Serial peripheral interface

ZRT Zero radius turning

References

[1] Analog Devices, Inc., AD7280 datasheet. http://www.analog.com/static/imported-files/data_sheets/AD7280.pdf.

[2] E. Bakker, L. Nyborg, and H. Pacejka, *Tyre Modelling for Use in Vehicle Dynamics Studies*, Society of Automotive Engineers, 870421, 1987.

[3] A. Bemporad and M. Morari, "Robust model predictivecontrol: A survey," *Robustness in Identification and Control*, pp. 207–226, 1999.

[4] H. Bergveld, W. Kruijt, P. Notten, and P. Notten. *Battery Management Systems: Design by Modelling*, Kluwer Academic Publishers, 2002.

[5] D. P. Bertsekas, "Constrained optimization and Lagrange multiplier methods," *Computer Science and Applied Mathematics*, vol. 1, 1982.

[6] L. Birgé and P. Massart, "From model selection to adaptive estimation. Festschrift for Lucien Le Cam," *Research Papers in Probability and Statistics*, pp. 55–87, 1997.

[7] B. Boada, M. Boada, and V. Diaz, "Fuzzy-logic applied to yaw moment control for vehicle stability," *Vehicle System Dynamics*, 43(10):753–770, 2005.

[8] B. K. Bose, *Modern Power Electronics and AC Drives*, Prentice Hall PTR, pp. 224–229, 2002.

[9] E. Camacho and C. Bordons, *Model Predictive Control*, Springer-Verlag, 2004.

[10] E. F. Camacho, C. Bordons, E. F. Camacho, and C. Bordons, *Model Predictive Control*, vol. 303, Springer, Berlin, 1999.

[11] M. Canale, L. Fagiano, A. Ferrara, and C. Vecchio, "Vehicle yaw control via second-order sliding-mode technique," *IEEE Transactions on Industrial Electronics*, 55(11):3908–3916, 2008.

[12] Chenyang, CYCT04-xnM20B datasheet. http://www.cy-sensors.com/CYCT04-xnM20B.pdf.

[13] D. Cho and J. K. Hedrick, "Automotive powertrain modeling for control," *Journal of Dynamic Systems, Measurement, and Control*, 111(4):568–576, 1989.

[14] M. Chow, S. Lukic, L. Wang, and A. Govindaraj, Research report on A123 battery modeling. http://www.adac.ncsu.edu/res_proj/battery_model/Docs/ResearchReportonA123BatteryModeling.ppt.

[15] M. Clerc, "The swarm and the queen: Towards a deterministic and adaptive particle swarm optimization," in *Proceedings of the Congress on Evolutionary Computation*, vol. 3, Institute of Electrical and Electronics Engineers, 1999.

[16] C. A. Coello and M. S. Lechuga, "MOPSO: A proposal for multiple objective particle swarm optimization," in *Proceedings of the Congress on Evolutionary Computation*, 2:1051–1056, Institute of Electrical and Electronics Engineers, 2002.

[17] K. Deb, *Multi-objective Optimization Using Evolutionary Algorithms*, Wiley, 2001.

[18] S. Dhameja, *Electric Vehicle Battery Systems*, Butterworth-Heinemann, Woburn, Mass., 2002.

[19] B. Divakar, K. Cheng, H. Wu, J. Xu, H. Ma, W. Ting, K. Ding, et al., "Battery management system and control strategy for hybrid and electric vehicle," in *International Conference on Power Electronics Systems and Applications*, pp. 1–6, 2009.

[20] R. Eberhart and J. Kennedy, "A new optimizer using particle swarm theory," in *Proceedings of the Sixth International Symposium on Micro Machine and Human Science*, pp. 39–43, Institute of Electrical and Electronics Engineers, 1995.

[21] Epcos, B82793 datasheet. http://www.epcos.com/inf/30/db/emc_00/02470251.pdf.

[22] E. Esmailzadeh, A. Goodarzi, and G. Vossoughi, "Directional stability and control of four-wheel independent drive electric vehicles," in *Proceedings of the Institution of Mechanical Engineers, Part K: Journal of Multi-body Dynamics*, 216(4):303–313, 2002.

[23] M. Ferdowsi, *Battery Charge Equalization—State of the Art and Future Trends*, Society of Automotive Engineers, 2005.

[24] Freescale, MC9S12XEP100 reference manual covers MC9S12XE family. http://www.freescale.com/files/microcontrollers/doc/data_sheet/MC9S12XEP100RMV1.pdf, 2010.

[25] H. Fujimoto, A. Tsumasaka, and T. Noguchi, "Direct yaw-moment control of electric vehicle based on cornering stiffness estimation," in *Proceedings of the 31st Annual Conference of IEEE Industrial Electronics Society*, pp. 2626–2630, Institute of Electrical and Electronics Engineers, 2005.

[26] S. Fujisawa, K. Ohkubo, T. Yoshida, N. Satonaka, Y. Shidama, and H. Yamaura, "Improved moving properties of an omnidirectional vehicle using stepping motor," in *Proceedings of the 36th IEEE Conference on Decision and Control*, 4:3654–3656, Institute of Electrical and Electronics Engineers, 1997.

[27] C. Geng, L. Mostefai, and Y. Hori, "Body slip angle observer for electric vehicle stability control based on empirical tire model with fuzzy logic approach," in *Proceedings of the 34th Annual Conference of IEEE Industrial Electronics*, pp. 108–113, Institute of Electrical and Electronics Engineers, 2008.

[28] G. Genta, "Motor vehicle dynamics: Modeling and simulation," vol. 43, World Scientific Publishing Company Incorporated, 1997.

[29] D. Goldberg and K. Deb, "A comparative analysis of selection schemes used in genetic algorithms," *Foundations of Genetic Algorithms*, 1:69–93, 1991.

[30] V. G. Gudise and G. K. Venayagamoorthy, "Evolving digital circuits using particle swarm," in *Proceedings of the International Joint Conference on Neural Networks*, 1:468–472, Institute of Electrical and Electronics Engineers, 2003.

[31] A. Hande and T. Stuart, "A selective equalizer for NiMH batteries," *Journal of Power Sources*, 138(1-2):327–339, 2004.

[32] B. Hassibi, A. Sayed, and T. Kailath, "Linear estimation in Krein spaces, Part II: Applications," *IEEE Transactions on Automatic Control*, 41(1):34–49, 1996.

[33] B. Hassibi, A. Sayed, and T. Kailath, "Linear estimation in Krein spaces, Part I: Theory," *IEEE Transactions on Automatic Control*, 41(1):18–33, 1996.

[34] P. He and Y. Hori, "Optimum traction force distribution for stability improvement of 4WD EV in critical driving condition," in *Proceedings of the IEEE International Workshop on Advanced Motion Control*, pp. 596–601, Institute of Electrical and Electronics Engineers, 2006.

[35] C.-L. Hwang and L.-J. Chang, "Trajectory tracking and obstacle avoidance of mobile robots in an intelligent space using mixed h_2/h_∞ decentralized control," *IEEE/ASME Transactions on Mechatronics*, 12(3):345–352, 2007.

[36] J. R. Belt, "Battery test manual for plug-in hybrid electric vehicles." Idaho National Laboratory (INL), 2010.

[37] Matlab Inc., *Virtual Reality User's Guide*, 2004.

[38] Matlab Inc., *Real-Time Windows Target User's Guide*, 2008.

[39] Infineon, TLE6250 datasheet. http://www.infineon.com/cms/en/product/findProductTypeByName.html?q=tle6250.

[40] H. Inoue, "Development of vehicle stability control system for active safety," *Journal of Society of Automotive Engineers of Japan*, 49:45–51, 1995.

[41] B. Jiang and B.Wang, "Parameter estimation of nonlinear system based on genetic algorithm," *Control Theory and Applications*, 17(1):150–152, 2000.

[42] V. Johnson, "Battery performance models in ADVISOR," *Journal of Power Sources*, 110(2):321–329, 2002.

[43] D. Karnopp, *Vehicle Stability*, vol. 171, Marcel Dekker, 2004.

[44] J. Kennedy and R. Eberhart, "Particle swarm optimization," in *Proceedings of the International Conference on Neural Networks*, 4:1942–1948, Institute of Electrical and Electronics Engineers, 1995.

[45] J. Kennedy and R. C. Eberhart, "A discrete binary version of the particle swarm algorithm," in *Proceedings of the World Multi-conference on Systemics, Cybernetics and Informatics*, 5:4104–4108, Institute of Electrical and Electronics Engineers, 1997.

[46] J. Kim and H. Kim, "Electric vehicle yaw rate control using independent in-wheel motor," in *Power Conversion Conference—Nagoya,*

2007, pp. 705–710, Institute of Electrical and Electronics Engineers, 2007.

[47] P. C. Krause, O. Wasynczuk, and S. D. Sudhoff, *Analysis of Electric Machinery*, McGraw-Hill, New York, 1986.

[48] K. Kutluay, Y. Cadirci, Y. Ozkazanc, and I. Cadirci, "A new online state-of-charge estimation and monitoring system for sealed lead-acid batteries in telecommunication power supplies," *IEEE Transactions on Industrial Electronics*, 52(5):1315–1327, 2005.

[49] T. L. Lam, Y. Xu, and G. Xu, "Traction force distribution on omni-directional four wheel independent drive electric vehicle," in *Proceedings of the IEEE International Conference on Robotics and Automation*, pp. 3724–3729, Institute of Electrical and Electronics Engineers, 2009.

[50] M. Lauria, I. Nadeau, P. Lepage, Y. Morin, P. Giguère, F. Gagnon, D. L'Etourneau, et al., "Design and control of a four steered wheeled mobile robot," in *Proceedings of the 32nd Annual Conference on Industrial Electronics*, pp. 4020–4025, Institute of Electrical and Electronics Engineers, 2006.

[51] D. Lee, S. Shiah, C. Lee, and Y. Wang, "State-of-charge estimation for electric scooters by using learning mechanisms," *IEEE Transactions on Vehicular Technology*, 56(2):544–556, 2007.

[52] Y. Lee and G. Cheng, "Quasi-resonant zero-current-switching bidirectional converter for battery equalization applications," *IEEE Transactions on Power Electronics*, 21(5):1213–1224, 2006.

[53] Y. Lee and M. Cheng, "Intelligent control battery equalization for series connected lithium-ion battery strings," *IEEE Transactions on Industrial Electronics*, 52(5):1297–1307, 2005.

[54] Y. Lee, M. Cheng, S. Yang, and C. Hsu, "Individual cell equalization for series connected lithium-ion batteries," *IEICE Transactions on Communications*, 89(9):2596–2607, 2006.

[55] Y. Lee, C. Duh, S. Yang, and G. Chen, "Battery equalization using bi-directional Cuk converter in DCVM operation," in *IEEE Power Electronics Specialists Conference*, 36:765–771, 2005.

[56] LEM, HTFS200-P/SP2 datasheet. http://www.lem.com/docs/products/htfs_sp2_e.pdf.

[57] B. Lindemark, "Individual cell voltage equalizers (ICE) for reliable battery performance," in *International Telecommunications Energy Conference*, pp. 196–201, 1991.

[58] J. M. Maciejowski, *Predictive Control: with Constraints*, Pearson Education, 2002.

[59] M. Makatchev, S. Y. Lang, S. Tso, and J. J. McPhee, "System design, modelling, and control of a four-wheel-steering mobile robot," in *Proceedings of the 19th Chinese Control Conference*, pp. 759–763, 2000.

[60] M. Makatchev, S. Tso, S. Lang, and J. J. McPhee, "Cross-coupling control for slippage minimization of a four-wheel-steering mobile robot," in *Proceedings of the International Symposium on Robotics*, 31:42–47, 2000.

[61] T. Markel, A. Brooker, T. Hendricks, V. Johnson, K. Kelly, B. Kramer, M. O'Keefe, et al., "ADVISOR: A systems analysis tool for advanced vehicle modeling," *Journal of Power Sources*, 110(2):255–266, 2002.

[62] P. J. Mckerrow and D. Ratner, "Calibrating a 4-wheel mobile robot," in *Proceedings of the 2002 IEEE/RSJ International Conference on Intelligent Robots and Systems*, 1:859–864. Institute of Electrical and Electronics Engineers, 2002.

[63] Microchip, MCP9701A datasheet. http://ww1.microchip.com/downloads/en/DeviceDoc/21942e.pdf.

[64] S. Mingli, C. Liang, G. Jianhua, and F. Yong, "Estimation of vehicle side slip angle using hybrid observer," in *Proceedings of the 29th Chinese Control Conference*, pp. 5378–5383, Institute of Electrical and Electronics Engineers, 2010.

[65] N. Miyamoto, M. Shino, Y. Wang, T. Kamada, and M. Nagai, "Robust traction control of EV considering vehicle stability," *Proceedings of Translog*, 99:231–234, 1999.

[66] S. Moore and P. Schneider, *A Review of Cell Equalization Methods for Lithium Ion and Lithium Polymer Battery Systems*, Society of Automotive Engineers, 2001.

[67] M. C. Nechyba and Y. Xu, "Cascade neural networks with node-decoupled extended Kalman filtering," in *Proceedings of the International Symposium on Computational Intelligence in Robotics and Automation*, pp. 214–219, Institute of Electrical and Electronics Engineers, 1997.

[68] M. C. Nechyba and Y. Xu, "Human control strategy: Abstraction, verification, and replication," *IEEE Control Systems Magazine*, 17(5):48–61, 1997.

[69] K. Ng, C. Moo, Y. Chen, and Y. Hsieh, "Enhanced coulomb counting method for estimating state-of-charge and state-of-health of lithium-ion batteries," *Applied Energy*, 86(9):1506–1511, 2009.

[70] P. Notten, J. Veld, and J. Beek, "Boostcharging Li-ion batteries: A challenging new charging concept," *Journal of Power Sources*, 145(1):89–94, 2005.

[71] Ministry of Industry and Information Technology of PRC, Investment guidance of technological upgrading and transformation in automotive industry. http://www.autoinfo.gov.cn/autoinfo_cn/xwzx/tbgz/webinfo/2010/05/1274923210160000.htm, 2010.

[72] Onsemi, NCV33152 datasheet. http://www.onsemi.com/pub_link/Collateral/MC34152-D.PDF.

[73] Onsemi, NUP2105L datasheet. http://www.onsemi.com/pub_link/Collateral/NUP2105L-D.pdf.

[74] R. P. Osborn and T. Shim, "Independent control of all-wheel-drive torque distribution," *Vehicle System Dynamics*, 44(7):529–546, 2006.

[75] H. Oshima, M. Tani, N. Kobayashi, A. Ishii, and K. Imai, "Control for four-wheel individual steering and four-wheel driven electronic vehicle," *Electrical Engineering in Japan*, 153(3):71–78, 2005.

[76] H. Pacejka, *Tyre and Vehicle Dynamics*, Society of Automotive Engineers/Butterworth-Heinemann, 2005.

[77] H.-S. Park, C.-E. Kim, C.-H. Kim, G.-W. Moon and J.-H. Lee, "A modularized charge equalizer for an HEV lithium-ion battery string," *IEEE Transaction on Industrial Electronics*, 56(5):1464–1476, 2009.

[78] H. Park, C. Kim, and G. Moon, "Charge equalizer design method based on battery modularization," in *IEEE International Conference on Sustainable Energy Technologies*, pp. 558–563, 2008.

[79] J. H. Park, "h_∞ Direct yaw-moment control with brakes for robust performance and stability of vehicles," *JSME International Journal Series C*, 44:404–413, 2001.

[80] K. Passino and S. Yurkovich, "Fuzzy control," *Citeseer*, 1998.

[81] P. Perona and J. Malik, "Scale-space and edge detection using anisotropic diffusion," *IEEE Transactions on Pattern Analysis and Machine Intelligence*, 12(7):629–639, 1990.

[82] A. Pesaran, "Battery thermal models for hybrid vehicle simulations," *Journal of Power Sources*, 110(2):377–382, 2002.

[83] D. W. Pi, N. Chen, B. J. Zhang, and G. H. Zhong, "Enhancements in vehicle stability with yaw moment control via differential braking," in *Proceedings of the IEEE International Conference on Vehicular Electronics and Safety*, pp. 136–140, 2009.

[84] G. Plett, "Extended Kalman filtering for battery management systems of LiPB-based HEV battery packs: Part 2. Modeling and identification," *Journal of Power Sources*, 134(2):262–276, 2004.

[85] G. Plett, "Extended Kalman filtering for battery management systems of LiPB-based HEV battery packs—Part 3, State and parameter estimation," *Journal of Power Sources*, 134(2):277–292, 2004.

[86] G. Plett, "Sigma-point Kalman filtering for battery management systems of LiPB-based HEV battery packs: Part 1: Introduction and state estimation," *Journal of Power Sources*, 161(2):1356–1368, 2006.

[87] G. Plett, "Sigma-point Kalman filtering for battery management systems of LiPB-based HEV battery packs: Part 2: Simultaneous state and parameter estimation," *Journal of Power Sources*, 161(2):1369–1384, 2006.

[88] G. Plett and M. Klein, "Advances in HEV battery management systems," in *Society of Automotive Engineers Congress*, CD-ROM Proceedings of SAE Convergence, 2006.

[89] J. Ploeg, A. C. van der Knaap, and D. J. Verburg, "Ats/agv-design, implementation and evaluation of a high performance agv," in *IEEE Intelligent Vehicle Symposium*, 1:127–134, Institute of Electrical and Electronics Engineers, 2002.

[90] H. Qian, T. L. Lam, W. Li, C. Xia, and Y. Xu, "System and design of an omni-directional vehicle," in *Proceedings of the IEEE International Conference on Robotics and Biomimetics*, pp. 389–394, Institute of Electrical and Electronics Engineers, 2009.

[91] S. Qin and T. Badgwell, "A survey of industrial model predictive control technology," *Control Engineering Practice*, 11(7):733–764, 2003.

[92] R. Rajamani, *Vehicle Dynamics and Control*, Springer, 2006.

[93] J. A. Rossiter, *Model-Based Predictive Control: A Practical Approach*, CRC, 2003.

[94] S. Sakai, "Lateral motion stabilization with feedback controlled wheels," in *Proceedings of the International Symposium on Advanced Vehicle Control*, pp. 815–820, 2002.

[95] S.-I. Sakai, H. Sado, and Y. Hori, "Dynamic driving/braking force distribution in electric vehicles with independently driven four wheels," *Electrical Engineering in Japan*, 138(1):79–89, 2002.

[96] Z. Salameh, M. Casacca, and W. Lynch, "A mathematical model for lead-acid batteries," *IEEE Transaction on Energy Conversion*, 7(1):93–98, 1992.

[97] S. Schroedl and W. Zhang, "Predicting driving speed using neural networks," in *Proceedings of the International Conference on Intelligent Transportation Systems*, 1:402–407, Institute of Electrical and Electronics Engineers, 2003.

[98] Y. Shi and R. Eberhart, "A modified particle swarm optimizer," in *Proceedings of the International Conference on Evolutionary Computation and the World Congress on Computational Intelligence*, pp. 69–73, Institute of Electrical and Electronics Engineers, 1998.

[99] M. Shino and M. Nagai, "Independent wheel torque control of small-scale electric vehicle for handling and stability improvement," *JSAE Review*, 24(4):449–456, 2003.

[100] D. Simon, *Optimal State Estimation: Kalman, H_∞ and Nonlinear Approaches*, Wiley, 2006.

[101] C. Speltino, D. Di Domenico, G. Fiengo, and A. G. Stefanopoulou, "Cell equalization in battery stacks through state of charge estimation polling," in *American Control Conference*, 2010.

[102] T. Stuart, F. Fang, X. Wang, C. Ashtiani, and A. Pesaran, "A modular battery management system for HEVs," In *Proceedings of the SAE Future Car Congress* (Paper Number 2002-01-1918), Arlington, VA, 2002.

[103] A. Szumanowski and Y. Chang, "Battery management system based on battery nonlinear dynamics modeling," *IEEE Transactions on Vehicular Technology*, 57(3):1425–1432, 2008.

[104] F. Tahami, R. Kazemi, and S. Farhanghi, "A novel driver assist stability system for all-wheel-drive electric vehicles," *IEEE Transactions on Vehicular Technology*, 52(3):683–692, 2003.

[105] M. Tang, T. Stuart, D. Co, and I. Kokomo, "Selective buck-boost equalizer for series battery packs," *IEEE Transactions on Aerospace and Electronic Systems*, 36(1):201–211, 2000.

[106] B. Thuilot, B. d"Andréa Novel, and A. Micaelli, "Modeling and feedback control of mobile robots equipped with several steering wheels," *IEEE Transactions on Robotics and Automation*, 12(3):375–390, 1996.

[107] Texas Instruments, TPS54160 datasheet. http://focus.ti.com/lit/ds/symlink/tps54160.pdf.

[108] I. C. Trelea, "The particle swarm optimization algorithm: Convergence analysis and parameter selection," *Information Processing Letters*, 85(6):317–325, 2003.

[109] H. E. Tseng, B. Ashrafi, D. Madau, T. Allen Brown, and D. Recker, "The development of vehicle stability control at Ford," *IEEE/ASME Transactions on Mechatronics*, 4(3):223–234, 1999.

[110] U.S. Energy Information Administration, "U.S. primary energy consumption by source and sector," 2008. http://www.eia.doe.gov/emeu/aer/pecss_diagram.html.

[111] F. van den Bergh and A. P. Engelbrecht, "Cooperative learning in neural networks using particle swarm optimizers," *South African Computer Journal*, pp. 84–90, 2000.

[112] K. K. Veeramachaneni, L. A. Osadciw, and P. K. Varshney, "Adaptive multimodal biometric fusion algorithm using particle swarm," in *Proceedings of SPIE*, 5099: 211, 2003.

[113] J. Wang, B. Cao, Q. Chen, and F. Wang, "Combined state of charge estimator for electric vehicle battery pack," *Control Engineering Practice*, 15(12):1569–1576, 2007.

[114] M. K. Yalçin, S. M. Yeşiloğlu, M. Dal, and H. Temeltaş, "Maneuvering strategies for four-wheel drive, four-wheel steer mobile robots using curvatures based on weingarten-maps," in *32nd Annual Conference on IEEE Industrial Electronics*, pp. 4148–4152, Institute of Electrical and Electronics Engineers, 2006.

[115] J. Yamakawa and K. Watanabe, "A method of optimal wheel torque determination for independent wheel drive vehicles," *Journal of Terramechanics*, 43(3):269–285, 2006.

[116] J. Yan, Q. Ling, and W. Chen, "Nonlinear model predictive control based on evolutionary algorithms: Framework, theory and application," in A. Giordano and G. Costa, eds., *Soft Computing: New Research*, Nova Science Publishers, 2008.

[117] J. Yan, G. Xu, Y. Xu, and B. Xie, "Battery state-of-charge estimation based on H_∞ filter for hybrid electric vehicle," in *International Conference on Control, Automation, Robotics and Vision*, pp. 464–469, 2008.

[118] K.-J. Yang, K.-S. Hong, and D.-I. Cho, "A robust control for engine and transmission systems: Enhancement of shift quality," *JSME International Journal Series C*, 44(3):697–707, 2001.

[119] H. Yoshida, K. Kawata, Y. Fukuyama, S. Takayama, and Y. Nakanishi, "A particle swarm optimization for reactive power and voltage control considering voltage security assessment," *IEEE Transactions on Power Systems*, 15(4):1232–1239, 2000.

[120] T. A. Stuart and W. Zhu, "Fast equalization for large lithium ion batteries," *IEEE Aerospace and Electronic Systems Magazine*, 24(7):27–31, 2009.

Index

R